NUMERICAL SIMULATION OF WAVES AND FRONTS IN INHOMOGENEOUS SOLIDS

WORLD SCIENTIFIC SERIES ON NONLINEAR SCIENCE

Editor: Leon O. Chua
University of California, Berkeley

*To view the complete list of the published volumes in the series, please visit:
http://www.worldscibooks.com/series/wssnsa_series.shtml

WORLD SCIENTIFIC SERIES ON
ONLINEAR SCIENCE

Series A   Vol. 62

 series Editor: Leon O. Chua

NUMERICAL SIMULATION OF WAVES AND FRONTS IN INHOMOGENEOUS SOLIDS

Arkadi Berezovski
Jüri Engelbrecht

Tallinn University of Technology, Estonia

Gérard A Maugin

Université Pierre et Marie Curie, France

World Scientific

NEW JERSEY · LONDON · SINGAPORE · BEIJING · SHANGHAI · HONG KONG · TAIPEI · CHENNAI

Published by

World Scientific Publishing Co. Pte. Ltd.

5 Toh Tuck Link, Singapore 596224

USA office: 27 Warren Street, Suite 401-402, Hackensack, NJ 07601

UK office: 57 Shelton Street, Covent Garden, London WC2H 9HE

Library of Congress Cataloging-in-Publication Data
Berezovski, Arkadi.
    Numerical simulation of waves and fronts in inhomogeneous solids / by Arkadi Berezovski,
Jüri Engelbrecht, Gérard A Maugin.
        p. cm. -- (World Scientific series on nonlinear science. Series A ; v. 62)
    Includes bibliographical references and index.
    ISBN-13: 978-981-283-267-2 (hardcover : alk. paper)
    ISBN-10: 981-283-267-X (hardcover : alk. paper)
    1. Elastic solids. 2. Inhomogeneous materials. 3. Wave motion, Theory of.
    I. Engelbrecht, Jüri. II. Maugin, G. A. (Gérard A.), 1944- III. Title.
    QA935.B375 2008
    530.4'12--dc22
                                                                    2008015670

British Library Cataloguing-in-Publication Data
A catalogue record for this book is available from the British Library.

Printed in Singapore.

Preface

Numerical simulations of dynamical problems in inhomogeneous solids are quite difficult for the case of moving discontinuities such as phase transition fronts or cracks. The origin of these difficulties is a constitutive deficiency in the thermomechanical description of the corresponding irreversible processes, which leads to an uncertainty in jump relations at moving discontinuities. Consequently, the construction of an appropriate numerical algorithm should be complemented by the development of the thermomechanical theory.

The aim of this book is to provide a framework for the description of moving discontinuities in solids and its implementation in a finite-volume numerical algorithm.

The macroscopic description of moving discontinuities in solids (such as phase-transition fronts or crack fronts) needs to go beyond the conventional local equilibrium approximation, since both phase-transition fronts and cracks propagate irreversibly accompanied by entropy production at their moving front.

The numerical simulation of thermomechanical processes supposes the application of a certain discretization method. Therefore, instead of infinitesimally small material "points", which are in local equilibrium, in numerical simulations we deal with finite-size computational cells, states of which are non-equilibrium ones in general. To keep the meaning of thermodynamic quantities well-defined, a projection of the non-equilibrium states onto an equilibrium subspace is used conventionally. This means that the non-equilibrium state of each computational cell is associated with an equilibrium state of the accompanying reversible process. The simplest projection is the averaging of corresponding fields over the computational cell. An averaging procedure over the control volume applied to appropriate conservation laws leads to a system of equations in terms of averaged quantities and corresponding fluxes at boundaries of cells. Such a procedure is a basis for Godunov-type finite-volume numerical schemes.

The main problem in the construction of a particular algorithm is the proper determination of numerical fluxes. However, even if we are able to determine the numerical fluxes in the bulk, their determination at moving fronts is highly questionable.

Throughout the book, all needed fluxes are determined by means of the local equilibrium jump relations. These local equilibrium jump relations are formulated in terms of excess quantities, which characterize the non-equilibrium states of computational cells. The continuity of excess quantities across the moving discontinuity is used for the closure of the model. This leads to the relation between the stress jump at the discontinuity and the corresponding driving force, which allows us to determine the velocity of a moving front. As a consequence, a thermodynamically consistent finite volume numerical algorithm is obtained, which can be applied to both wave and front propagation problems.

It is remarkable that similar considerations are applicable to another problem of moving discontinuity, namely, to crack dynamics. In the case of a moving crack, the local equilibrium jump relation at the crack front is complemented by another assumption regarding the excess quantities, which reflects the distinction between the problems of moving phase boundaries and cracks.

Examples of numerical simulations of phase-transition front propagation and straight brittle crack motion show the applicability of the developed non-equilibrium description to solve the problems of moving fronts.

A. Berezovski, J. Engelbrecht, G.A. Maugin

Contents

Chapter 1

Introduction

In this introductory chapter we explain first the main notions used in the book: waves, fronts, inhomogeneities, etc. and present some instructive examples which illustrate the topics. Then the outline of the book is described: theoretical considerations in the first part, followed by the analysis of one- and two-dimensional problems.

1.1 Waves and fronts

Waves and fronts are apparently similar. However, there are slight but essential differences between them in the admitted classical view. If waves essentially correspond to continuous variations of the states of material points representing a medium, then fronts are discontinuity surfaces (or lines) dividing the medium into distinct parts. But the shared characteristic feature of waves and fronts is their motion. The motion of waves and fronts is governed by the same governing equations, namely, by the fundamental conservation laws for mass, linear momentum, and energy in the standard cases. This is why we consider wave and front propagations under a single umbrella.

However, if these conservation laws (complemented by constitutive relations) are sufficient for the description of thermoelastic waves [Achenbach (1973); Graff (1975); Bedford and Drumheller (1994); Billingham and King (2000)], then for cracks or phase-transition fronts we need more. Certain additional relations are required in the presence of moving discontinuities in order to isolate a unique solution amongst the whole set of possible solutions satisfied by given initial and boundary conditions.

1

1.2   True and quasi-inhomogeneities

Our attention is concentrated on the wave and front propagation in inhomogeneous media. Material inhomogeneities are theoretically seen as defects in the translational invariance of material properties on the material manifold. Amongst these, we find *true material inhomogeneities*, e.g. material regions of rapid changes in material properties due to a change of constitution; these changes may be more or less smooth or even abrupt. We also find physical and field properties which manifest themselves as *quasi-inhomogeneities* in a general theory of inhomogeneity [Maugin (2003)], i.e. they contribute in the same way as true inhomogeneities to the so-called balance of linear momentum on the material manifold, an equation that just built for that purpose. If the location of rapid changes in material properties due to a change of constitution (true inhomogeneities) is prescribed as initial data, then in contrast quasi-inhomogeneities, such as cracks or phase-transition fronts, are moved during the process and are therefore part of the solution of a particular problem. This is what make this problem simultaneously more interesting and more challenging.

1.3   Driving force and the corresponding dissipation

Here quasi-inhomogeneities provided by field singularities are considered. Unless they have a fractal structure, the support of these singularities may only be zero-, one- or two-dimensional in our three-dimensional physical and material spaces. We focus here on one- and two-dimensional singular sets. As these sets appear to be displaced as a consequence of the general evolution of a field solution under the time-dependent boundary conditions, driving forces acting on them are defined by duality with velocities of points of the sets in agreement with the basic vision of mechanics. The power expended by the driving force should finally be written as the general bilinear form

$$P(\mathbf{f}) = \mathbf{f} \cdot \mathbf{V}, \tag{1.1}$$

where $\mathbf{f}$ is the driving force, and $\mathbf{V}$ is the material velocity of points of the set [Maugin (1993)]. In some cases, the observed motion of singular sets is thermodynamically irreversible, and the force $\mathbf{f}$ of a non-Newtonian nature acquires a physical meaning only through the power (Eq. (1.1)) it expends, as this is, in fact, its definition in a weak mathematical formulation on the material manifold. The irreversible progress of the singularity set is then governed by the second law of thermodynamics. This means (in terms of temperature θ_S and entropy production σ_S at the singularity) that

$$\mathbf{f} \cdot \mathbf{V} = \theta_S \, \sigma_S \geq 0, \tag{1.2}$$

and the closure of the full solution of the evolution problem requires the formulation of a kinetic law relating $\mathbf{f}$ and $\mathbf{V}$ or a hypothesis about entropy production at the singularity. Examples of singular material sets exhibiting a driving force and dissipation are provided by cracks and phase-transition fronts, two examples that will recur in this book.

1.4 Example of a straight brittle crack

Consider a regular (simply connected) material body in which a straight through crack $C(t)$ expands with material velocity V_C at the crack tip (Fig. 1.1). The extension of the crack is collinear to the crack. The crack

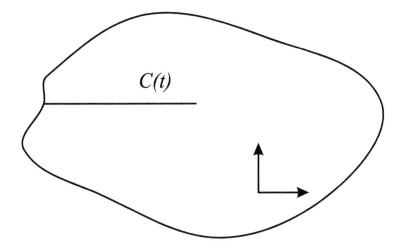

Fig. 1.1 Macroscopic straight-through crack.

tip, in fact, is a straight line but it is seen as a point in a sectional figure drawn orthogonally to this line.

A so-called material force acting at the crack tip (sucking the crack in the body) is none other than the energy-release rate G [Maugin (2000)]

$$G = FV_C = \theta_C \, \sigma_C \geq 0. \tag{1.3}$$

Here the subscript C denotes the crack tip, σ_C is the corresponding entropy production, and θ_C is the temperature at the crack tip.

It should be noted that the energy-release (*time*) rate G may be defined as the decrease of total potential energy Π due to a motion *in a time unit*

$$G = -\frac{\partial \Pi}{\partial t}. \tag{1.4}$$

According to Dascalu and Maugin (1993), this is nothing but the *dissipated power*

$$P_{diss} = G = \mathbf{f} \cdot \mathbf{V} \geq 0, \tag{1.5}$$

where, as in Eq. (1.1), $\mathbf{f}$ is the material driving force, and $\mathbf{V}$ is the material velocity of the crack tip.

Another definition of the energy-release rate (without time involved) is given by (cf. [Ravi-Chandar (2004)])

$$G^* = -\frac{\partial \Pi}{\partial a}, \tag{1.6}$$

where a is homogeneous to a length, e.g., displacement of the crack tip (increase in length) so that Eq. (1.6) is in fact a *material gradient, hence a material force*. Of course one is tempted to write

$$G = G^* \dot{a} = \left(-\frac{\partial \Pi}{\partial a} \right) \dot{a}, \tag{1.7}$$

a result that looks like Eq. (1.5) or Eq. (1.3).

In quasi-statics and in the absence of thermal and intrinsic dissipations, the energy-release rate can be computed by means of the celebrated *J*-integral of fracture [Rice (1968)] that is known to be path-independent and, therefore, provides a very convenient estimation tool once the field solution is known. However, the velocity at the crack tip remains undetermined and requires additional consideration.

1.5   Example of a phase-transition front

A similar situation holds for a displacive phase-transition front propagation. A stress-induced martensitic phase transformation in a single crystal of a thermoelastic material occurs by the fast propagation of sharp interfaces through the material (Fig. 1.2).

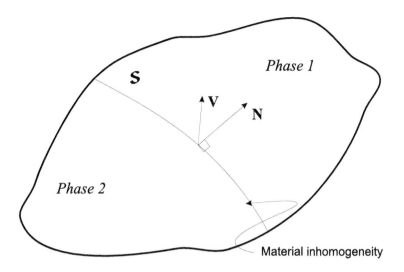

Fig. 1.2   Discontinuity front.

Following the already mentioned balance of pseudomomentum [Maugin (1993)], one can determine the driving force acting on the interface between phases. However, as in the case of the crack tip, this is not enough to determine the velocity of the phase-transition front in crystalline solids. Additional constitutive information is usually provided in the form of a kinetic relation between the driving force and the velocity of the phase boundary. The notion of a kinetic relation is introduced by Abeyaratne and Knowles (1991) following ideas from materials science. It is well understood [Abeyaratne, Bhattacharya and Knowles (2001)] that the kinetic relation is required because of the non-equilibrium character of the phase transformation process. The derivation of the kinetic relation is therefore the main problem in the description of the moving front propagation.

1.6   Numerical simulations of moving discontinuities

To perform numerical simulations of the motion of a discontinuity front we need to know the answers to the following questions in addition to the routine tasks of mesh generation and discretization of governing equations:

- how to compute the value of the driving force acting at the front (and its critical value);
- how to calculate the velocity of the front;

• how to determine the jumps of fields across the discontinuity front.

As we will see, the answers to the formulated questions can be given by a combination of the material formulation of continuum mechanics, the thermodynamics of discrete systems, and finite-volume numerical methods. The description of the corresponding framework and its application is in fact the main subject of the book.

It should be noted that the macroscopic description of moving fronts in solids (such as phase-transition fronts or crack fronts) is based on the conventional local equilibrium approximation (see, e.g. [Freund (1990); Ravi-Chandar (2004)] for cracks and [Abeyaratne and Knowles (2006)] for phase-transition fronts). This means that all the thermodynamic quantities, including temperature and entropy are defined following conventional methods [Callen (1960)]. Extension of the local equilibrium approximation to computational cells of finite volume computational schemes is not really straightforward. First, the local equilibrium values of field variables may be, in general, distinct from their averaged values, as shown by Muschik and Berezovski (2004). Secondly, the numerical fluxes needed for the updating of the local equilibrium state in a computational cell cannot be provided at a moving discontinuity without taking into account the entropy production due to the motion of this discontinuity.

Fortunately, there is a rather nice similarity between the discrete representation of conservation laws [LeVeque (2002a)] and the thermodynamics of discrete systems [Muschik (1993)]. Using this similarity, the Godunov-type numerical schemes for simulation of wave and front propagation can be reformulated in terms of excess quantities [Berezovski and Maugin (2001, 2002)], which appear in the local equilibrium approximation due to the interaction between discrete systems. Moreover, numerical fluxes are determined by means of the non-equilibrium jump relations at moving fronts [Berezovski and Maugin (2005b)], which are also formulated in terms of excess quantities [Berezovski and Maugin (2004)]. The closure of the model is achieved by an assumption concerning the excess quantities behavior across the moving front. This leads to the relation between the stress jump at the discontinuity and the corresponding driving force. Then the velocity of a moving front (the kinetic relation) can be determined [Berezovski and Maugin (2005a)]. As a consequence, we obtain a thermodynamically consistent numerical algorithm which can be applied to both wave and front propagation problems.

1.7   Outline of the book

The whole book is virtually divided into three parts. The first part includes the introduction and three theoretical chapters. In the second chapter, we briefly recall the description of inhomogeneities in the framework of the *material formulation* of continuum mechanics [Maugin (1993); Kienzler and Herrmann (2000)]. The third chapter is devoted to the derivation of non-equilibrium jump relations at discontinuities. The small-strain approximation of balance laws and jump relations used in applications is given in the fourth chapter. The general description of the applied finite volume algorithm is also presented.

In the second part of the book, the developed theory is applied to one-dimensional problems of wave and front propagation. In the fifth chapter, the application of the composite wave-propagation algorithm to the solution of wave propagation problems in media with rapidly-varying properties is demonstrated. The main advantage of the proposed theory is the possibility of its extension to the case of moving boundaries in solids. In this connection, a non-equilibrium description of martensitic phase-transition front propagation in solids is considered in the sixth chapter. The results of numerical simulations of moving phase boundaries in the one-dimensional setting are presented and the comparison with experimental data for impact-induced martensitic phase transition is also given.

Two-dimensional problems are considered in the third part of the book. Wave propagation in heterogeneous solids is the subject of the seventh chapter. Details of the numerical algorithm in two dimensions are also given here. The next chapter is devoted to waves in functionally graded materials. Examples of two-dimensional phase-transition fronts dynamics are considered in the ninth chapter. In the tenth chapter, the dynamics of a straight brittle crack is studied. The velocity of the crack is determined in terms of the driving force by means of non-equilibrium jump relations at the crack front. Numerical simulations are compared with the corresponding experimental data.

Overall conclusions are given in the last chapter. Details of the thermodynamic description in the framework of the thermodynamics of discrete systems are presented in the Appendix.

Chapter 2

Material Inhomogeneities in Thermomechanics

The classical theories of continua including the theory of thermoelasticity describe the behavior of homogeneous bodies. However, in reality the assumption of homogeneity is not sufficient: materials may be only piecewise homogeneous, or may have regions with rapidly changing properties or have microstructure. In such cases which are actually typical in contemporary technology, the theories should be generalized in order to grasp complicated structural and/or kinematical properties. In this chapter we explain basic theoretical considerations in order to be prepared for more detailed analysis in the next chapters.

Classical linear thermoelasticity is presented in various standard texts such as those by Boley and Weiner (1960); Carlson (1972); Hetnarski (1986) and Nowacki (1986). For the nonlinear framework, we refer the reader to Suhubi (1975); Maugin (1988). Here we give the relevant field equations and constitutive equations at any regular material point $\mathbf{X}$ in a material body. The *Piola-Kirchhoff formulation* [Maugin (1993)] is used from the start. This greatly simplifies the form of *jump relations* associated with the conservation laws. A direct (no index) notation is used whenever there is no ambiguity. An index (Cartesian tensor) notation may be re-introduced where we feel it necessary or more convenient.

2.1 Kinematics

The material body is considered an open, simply connected subset $\mathcal{B}$ of the material manifold $\mathcal{M}^3$ of material points (or particles P), which are referred to a position $\mathbf{X}$ in a reference configuration κ_0 (Fig. 2.1).

Let κ_t be the actual (at time t) configuration of the solid body $\mathcal{B}$ in physical space E^3. The direct time-parametrized motion of $\mathbf{X}$ is given by

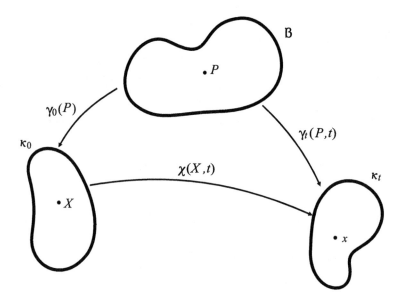

Fig. 2.1   Body and configurations.

the sufficiently regular function

$$\mathbf{x} = \chi(\mathbf{X}, t), \qquad (2.1)$$

which represents the time sequence of physical configurations occupied by the material point $\mathbf{X}$ in E^3 as time goes on. The physical velocity $\mathbf{v}$ and the direct-motion gradient $\mathbf{F}$ are defined by

$$\mathbf{v} := \left.\frac{\partial \chi}{\partial t}\right|_{\mathbf{X}}, \qquad \mathbf{F} := \left.\frac{\partial \chi}{\partial \mathbf{X}}\right|_{t} \equiv \nabla_R \chi. \qquad (2.2)$$

In terms of Cartesian components, the deformation gradient is represented by

$$F^i{}_K = \frac{\partial x^i}{\partial X^K}. \qquad (2.3)$$

As concerns the inverse-motion description, it is assumed that

$$J_F = \det \mathbf{F} > 0 \qquad (2.4)$$

always. Then the inverse motion

$$\mathbf{X} = \chi^{-1}(\mathbf{x}, t) \qquad (2.5)$$

is a well-behaved function. Thus, the material velocity $\mathbf{V}$ and the inverse-motion gradient $\mathbf{F}^{-1}$ can be defined by [Maugin (1993)]

$$\mathbf{V} := \left.\frac{\partial \chi^{-1}}{\partial t}\right|_{\mathbf{x}}, \qquad \mathbf{F}^{-1} := \left.\frac{\partial \chi^{-1}}{\partial \mathbf{x}}\right|_{t}. \tag{2.6}$$

At all regular material points, $\mathbf{v}$ and $\mathbf{V}$ are shown to be related by

$$\mathbf{v} + \mathbf{F} \cdot \mathbf{V} = 0. \tag{2.7}$$

This is proved via the chain rule of differentiation [Maugin (1993)]. Here the operation of a tensor on a vector (dotting a tensor from the right by a vector) is defined as follows:

$$\mathbf{v} = -\mathbf{F} \cdot \mathbf{V} \quad \equiv \quad v^{i} = -F^{i}{}_{K} V^{K}. \tag{2.8}$$

2.2 Integral balance laws

We now consider the case of thermoelastic conductors of heat. The integral balance laws can be written using the configuration κ_0, referring to the material body $\mathcal{B}$ and its bounding surface $\partial\mathcal{B}$ of unit outward normal $\mathbf{N}$ in the absence of external force and heat supply, as follows [Maugin (1992, 1993)]:

Balance of mass

$$\frac{d}{dt} \int_{\mathcal{B}} \rho_0(\mathbf{X}) \, dV = 0; \tag{2.9}$$

Balance of linear momentum

$$\frac{d}{dt} \int_{\mathcal{B}} \mathbf{p} \, dV = \int_{\partial\mathcal{B}} \mathbf{N} \cdot \mathbf{T} \, dA; \tag{2.10}$$

Balance of energy

$$\frac{d}{dt} \int_{\mathcal{B}} \mathcal{H} \, dV = \int_{\partial\mathcal{B}} \mathbf{N} \cdot (\mathbf{T} \cdot \mathbf{v} - \mathbf{Q}) \, dA; \tag{2.11}$$

Entropy inequality

$$\frac{d}{dt} \int_{\mathcal{B}} S \, dV + \int_{\partial\mathcal{B}} \mathbf{N} \cdot \mathbf{S} \, dA \geq 0, \qquad \mathbf{S} = \mathbf{Q}/\theta, \tag{2.12}$$

where t is time, $\rho_0(\mathbf{X})$ is the matter density in the reference configuration, $\mathbf{p} = \rho_0 \mathbf{v}$ is the linear momentum, $\mathbf{T}$ is the first Piola-Kirchhoff stress tensor, $\mathcal{H} = K + E$, $K = \frac{1}{2}\rho_0 \mathbf{v}^2$ is the kinetic energy per unit volume in the reference configuration, E is the corresponding internal energy, S is the entropy per unit volume, $\theta > 0$ is the temperature (inf $\theta = 0$), $\mathbf{Q}$ is the material heat flux, and $\mathbf{S}$ is the material entropy flux. Dotting a tensor from the left by a vector is defined by

$$(\mathbf{N} \cdot \mathbf{T})_i \equiv N_K T^K_{\ i}. \qquad (2.13)$$

2.3 Localization and jump relations

Let $\mathcal{A}(\mathbf{X}, t)$ be a material vector field which is only piecewise continuous over the material body $\mathcal{B}$ of regular boundary $\partial\mathcal{B}$. A smooth discontinuity surface such as $\mathcal{S}$ moving at material velocity $\bar{\mathbf{V}}$, divides $\mathcal{B}$ into two regular subregions $\mathcal{B}^\pm$. Then we have the following theorems generalizing the Green-Gauss theorem and the Reynolds transport theorem [Maugin (1999a)]:

$$\int_{\mathcal{B}} \nabla_R \cdot \mathcal{A} \, dV = \int_{\partial\mathcal{B} - \mathcal{S}} \mathbf{N} \cdot \mathcal{A} \, dA + \int_{\mathcal{S}} \mathbf{N} \cdot [\mathcal{A}] \, dA, \qquad (2.14)$$

$$\int_{\mathcal{B}} \left. \frac{\partial \mathcal{A}}{\partial t} \right|_{\mathbf{X}} dV = \frac{d}{dt} \int_{\mathcal{B} - \mathcal{S}} \mathcal{A} \, dV - \int_{\mathcal{S}} \bar{V}_N [\mathcal{A}] \, dA, \qquad (2.15)$$

where $\bar{V}_N$ is the normal component of the material velocity of the points of $\mathcal{S}$. Here $[\mathcal{A}] = \mathcal{A}^+ - \mathcal{A}^-$, and $\mathcal{A}^\pm$ are the uniform limits of the field $\mathcal{A}$ in approaching the front from its positive and negative sides, respectively, along the unit normal $\mathbf{N}$ to the surface oriented from its negative to its positive side. Applying these theorems to integral balance laws we can obtain both local balance laws and jump relations.

2.3.1 *Local balance laws*

By localization, at any regular point $\mathbf{X}$ outside $\mathcal{S}$, we then have the following *local balance laws* [Maugin (1997, 1998)]

$$\left. \frac{\partial}{\partial t} \rho_0 \right|_{\mathbf{X}} = 0, \qquad (2.16)$$

$$\left. \frac{\partial}{\partial t} \mathbf{p} \right|_{\mathbf{X}} - div_R \mathbf{T} = 0, \qquad (2.17)$$

$$\left. \frac{\partial}{\partial t} \mathcal{H} \right|_{\mathbf{X}} - \nabla_R \cdot (\mathbf{T} \cdot \mathbf{v} - \mathbf{Q}) = 0, \qquad (2.18)$$

and the inequality of Clausius-Duhem

$$\left.\frac{\partial S}{\partial t}\right|_{\mathbf{X}} + \nabla_R \cdot \mathbf{S} \geq 0. \tag{2.19}$$

Here the divergence of a second-order tensor field is the vector with components

$$(div_R \mathbf{T})_i \equiv \frac{\partial T^K_{\ i}}{\partial X^K}, \tag{2.20}$$

and the divergence of a vector field is calculated as usual:

$$div_R \mathbf{Q} = \nabla_R \cdot \mathbf{Q} \equiv \frac{\partial Q^K}{\partial X^K}. \tag{2.21}$$

The energy equation (2.18) in the bulk can also be written in the form of the *heat propagation* equation ultimately governing the temperature

$$\left.\theta \frac{\partial S}{\partial t}\right|_{\mathbf{X}} + \nabla_R \cdot \mathbf{Q} = 0, \tag{2.22}$$

because there is neither heat body source nor intrinsic dissipation.

2.3.2 *Jump relations*

Remaining contributions to the integral balance laws (2.9) - (2.12) form the corresponding *jump relations* across $\mathcal{S}$. These are called the *Rankine-Hugoniot* jump relations [Maugin (1997, 1998)]

$$\bar{V}_N [\rho_0] = 0, \tag{2.23}$$

$$\bar{V}_N [\mathbf{p}] + \mathbf{N} \cdot [\mathbf{T}] = 0, \tag{2.24}$$

$$\bar{V}_N [\mathcal{H}] + \mathbf{N} \cdot [\mathbf{T} \cdot \mathbf{v} - \mathbf{Q}] = 0, \tag{2.25}$$

and

$$\bar{V}_N [S] - \mathbf{N} \cdot [\mathbf{Q}/\theta] = \sigma_{\mathcal{S}} \geq 0. \tag{2.26}$$

2.3.3   *Constitutive relations*

In the classical thermoelasticity of conductors, the constitutive equations (laws of state) are given in terms of free energy per unit volume, $W = W(\mathbf{F}, \theta)$, by

$$\mathbf{T} = \frac{\partial W}{\partial \mathbf{F}}, \qquad S = -\frac{\partial W}{\partial \theta}, \tag{2.27}$$

and Eq. (2.19) can be looked upon as a mathematical constraint on the formulation of the constitutive equation for entropy flux. This flux should also satisfy a "continuity" requirement such that

$$\mathbf{S}(\mathbf{F}, \theta, \nabla_R \theta) \to 0 \quad \text{as} \quad \nabla_R \theta \to 0. \tag{2.28}$$

Equations (2.16) to (2.28), together with a more precise expression for W and some of its mathematical properties (e.g., convexity), are those to be used in studying sufficiently regular nonlinear dynamical processes in thermoelastic conductors.

Here the definition for the derivative of a function of a tensor variable in terms of Cartesian components is represented as

$$T^K_{\ i} = \frac{\partial W}{\partial F^i_{\ K}}. \tag{2.29}$$

2.4   True and quasi-material inhomogeneities

The classical balance of linear momentum (with components in *physical* space) (2.17) is related to the invariance under space translations in that space (invariance under *spatial* translations) in the absence of body forces (e.g., gravity which acts at the current physical point or *placement* x). This is called *homogeneity of space* [Landau and Lifshitz (1986)]. Material space, i.e. matter per se, has no reason to be homogeneous. On the contrary, many materials are inhomogeneous in a smooth or more irregular or singular way.

Material inhomogeneities are, in general, defects in the translational invariance of material properties on the material manifold. Among these we find true material inhomogeneities, e.g. material regions of rapid changes in material properties due to a change of constitution; these changes may be more or less smooth or even abrupt. We also find physical and field properties which manifest themselves as quasi-inhomogeneities, i.e. they contribute in the same way as true inhomogeneities to the balance of linear momentum on the material manifold.

Following Maugin (2003), we call *quasi- (or pseudo-) inhomogeneity effects* in continuum mechanics those mechanical effects of any origin which manifest themselves as so-called *material forces* in the material mechanics of materials [Maugin (1993, 1995)]. The reason for these is that the force exerted on a true material inhomogeneity (a region of a material body where material properties vary with the material point or are different from those at other points outside the region) in a material displacement (caused by the field solution of the problem) is through the inherent duality of continuum mechanics the best characterization of the material inhomogeneity of a body. Forces acting on smooth distributions of dislocations (one kind of crystalline defect) and forces acting on macroscopic defects viewed as field singularities of certain dimensions on the material manifold, such as the forces driving macroscopic cracks or phase-transition fronts, are of this type.

However, the above-mentioned formalism does not reveal so far any material inhomogeneities. That is, even if, more generally than Eq. (2.27), we consider a free energy such as

$$W = W(\mathbf{F}, \theta, \mathbf{X}), \tag{2.30}$$

with an explicit dependence of free energy on the material point $\mathbf{X}$, the preceding set of equations is not formally modified, so this is not an intrinsic formulation. We need to write an equation of balance directly on the material manifold to see a manifestation of that inhomogeneity. This is obtained by projecting canonically Eq. (2.17) onto the material manifold $\mathcal{M}^3$ of points $\mathbf{X}$ constituting the body [Maugin and Trimarco (1995); Maugin (1997, 1998)].

2.4.1 *Balance of pseudomomentum*

The above-mentioned projection of the balance of linear momentum onto the material manifold can be achieved as follows. Multiplying Eq. (2.17) by $\mathbf{F}$ on the right we have

$$\left.\frac{\partial}{\partial t}(\rho_0 \mathbf{v})\right|_{\mathbf{X}} \cdot \mathbf{F} - (div_R \mathbf{T}) \cdot \mathbf{F} = \mathbf{0}. \tag{2.31}$$

The last equation can be rewritten as follows

$$\left.\frac{\partial}{\partial t}(\rho_0 \mathbf{v} \cdot \mathbf{F})\right|_{\mathbf{X}} - \rho_0 \mathbf{v} \cdot \left.\frac{\partial \mathbf{F}}{\partial t}\right|_{\mathbf{X}} - div_R(\mathbf{T} \cdot \mathbf{F}) + \mathbf{T} : (\nabla_R \mathbf{F})^T = \mathbf{0}, \tag{2.32}$$

since (superscript T denotes transposition)

$$(div_R \mathbf{T}) \cdot \mathbf{F} = div_R(\mathbf{T} \cdot \mathbf{F}) - \mathbf{T} : (\nabla_R \mathbf{F})^T. \tag{2.33}$$

In terms of components, the former equation is represented as

$$
\begin{aligned}
\frac{\partial T^K_{\ i}}{\partial X^K} F^i_{\ L} &= \frac{\partial}{\partial X^K}(T^K_{\ i} F^i_{\ L}) - T^K_{\ i}\frac{\partial F^i_{\ L}}{\partial X^K} \\
&= \frac{\partial}{\partial X^K}(T^K_{\ i} F^i_{\ L}) - T^K_{\ i}\frac{\partial F^i_{\ K}}{\partial X^L}.
\end{aligned} \tag{2.34}
$$

We consider the second and fourth terms on the left-hand side of Eq. (2.32) in more detail. For the second term we will have

$$\rho_0 \mathbf{v} \cdot \left.\frac{\partial \mathbf{F}}{\partial t}\right|_{\mathbf{X}} = \rho_0 \mathbf{v} \cdot (\nabla_R \mathbf{v})^T, \tag{2.35}$$

because the material time derivative of the direct-motion gradient tensor can be represented in the component form as follows

$$\left.\frac{\partial F^i_{\ K}}{\partial t}\right|_{\mathbf{X}} = \left.\frac{\partial}{\partial t}\left(\frac{\partial x^i}{\partial X^K}\right)\right|_{\mathbf{X}} = \left.\frac{\partial}{\partial X^K}\left(\frac{\partial x^i}{\partial t}\right)\right|_{\mathbf{X}} = \frac{\partial v^i}{\partial X^K}. \tag{2.36}$$

Arranging the right-hand side of Eq. (2.35) as

$$
\begin{aligned}
\rho_0 \mathbf{v} \cdot (\nabla_R \mathbf{v})^T &= \nabla_R\left(\frac{1}{2}\rho_0 \mathbf{v}^2\right) - \frac{1}{2}\mathbf{v}^2 \nabla_R \rho_0 \\
&= div_R\left(\frac{1}{2}\rho_0 \mathbf{v}^2\right)\mathbf{I} - \frac{1}{2}\mathbf{v}^2 \nabla_R \rho_0,
\end{aligned} \tag{2.37}
$$

we finally obtain

$$\rho_0 \mathbf{v} \cdot \left.\frac{\partial \mathbf{F}}{\partial t}\right|_{\mathbf{X}} = div_R\left(\frac{1}{2}\rho_0 \mathbf{v}^2\right)\mathbf{I} - \frac{1}{2}\mathbf{v}^2 \nabla_R \rho_0. \tag{2.38}$$

To represent the fourth term in Eq. (2.32), we consider the material gradient of the free energy

$$
\begin{aligned}
\nabla_R W(\mathbf{F}, \theta, \mathbf{X}) &= div_R(W\mathbf{I}) \\
&= (\nabla_R W)_{expl} + \frac{\partial W}{\partial \mathbf{F}} : (\nabla_R \mathbf{F})^T + \frac{\partial W}{\partial \theta}\nabla_R \theta,
\end{aligned} \tag{2.39}
$$

or, on account of the constitutive Eqs. (2.27),

$$div_R(W\mathbf{I}) = (\nabla_R W)_{expl} + \mathbf{T} : (\nabla_R \mathbf{F})^T - S\nabla_R \theta. \tag{2.40}$$

Here the following notation is used

$$(\nabla_R W)_{expl} = \left.\frac{\partial W}{\partial \mathbf{X}}\right|_{fixed\ fields}. \tag{2.41}$$

Therefore, the fourth term in Eq. (2.32) can be represented as follows

$$\mathbf{T} : (\nabla_R \mathbf{F})^T = div_R(W\mathbf{I}) - (\nabla_R W)_{expl} + S\nabla_R \theta. \qquad (2.42)$$

Substituting the obtained relations (2.38) and (2.42) into Eq. (2.32), we will finally have

$$\frac{\partial}{\partial t}(\rho_0 \mathbf{v} \cdot \mathbf{F})\Big|_{\mathbf{X}} + div_R \left((W - \frac{1}{2}\rho_0 \mathbf{v}^2)\mathbf{I} - \mathbf{T} \cdot \mathbf{F} \right)$$
$$+ \frac{1}{2}\mathbf{v}^2 \nabla_R \rho_0 + S\nabla_R \theta - (\nabla_R W)_{expl} = \mathbf{0}. \qquad (2.43)$$

The last equation is the desired projection of the balance of linear momentum onto the material manifold. In fact, defining the Lagrangian density of energy per unit volume $\mathcal{L}$

$$\mathcal{L} := \frac{1}{2}\rho_0(\mathbf{X})\mathbf{v}^2 - W(\mathbf{F}, \theta, \mathbf{X}), \qquad (2.44)$$

the dynamical Eshelby stress tensor $\mathbf{b}$

$$\mathbf{b} := -(\mathcal{L}\mathbf{I} + \mathbf{T} \cdot \mathbf{F}), \qquad (2.45)$$

the pseudomomentum $\mathcal{P}$

$$\mathcal{P} := -\rho_0 \mathbf{v} \cdot \mathbf{F}, \qquad (2.46)$$

and the inhomogeneity forces

$$\mathbf{f}^{inh} := \frac{1}{2}\mathbf{v}^2 \nabla_R \rho_0 - (\nabla_R W)_{expl}, \quad \mathbf{f}^{th} := S\nabla_R \theta, \qquad (2.47)$$

we can rewrite Eq. (2.43) as the following fully material balance of momentum called the *balance of pseudomomentum* in continuum physics:

$$\frac{\partial \mathcal{P}}{\partial t}\Big|_{\mathbf{X}} - div_R \mathbf{b} = \mathbf{f}^{inh} + \mathbf{f}^{th}. \qquad (2.48)$$

As we have seen, at all regular material points $\mathbf{X}$ Eq. (2.48) is a differential identity deduced from Eq. (2.17). An embryonic form of this equation for the case of statics in a purely hyperelastic homogeneous body in the absence of applied force may be found in Ericksen (1977) – we referred to this as Ericksen's identity [Maugin (1993)].

According to Eq. (2.47) the "force" $\mathbf{f}^{inh}$ indeed captures the explicit $\mathbf{X}$-dependency and deserves its naming as *material force of inhomogeneity*, or for short *inhomogeneity* force. This is the first cause for the momentum equation (2.48) to be *inhomogeneous* (i.e. to have a source term) while the original – in physical space – momentum equation (2.17) was a strict conservation law. Even more surprising is that a spatially nonuniform state of temperature ($\nabla_R \theta \neq \mathbf{0}$) causes a similar effect, i.e. the material *thermal* force $\mathbf{f}^{th}$ acts just like a true material inhomogeneity in so far as the balance of canonical (material) momentum is concerned [Epstein and Maugin (1995)].

The energy equation (2.18) and the balance of pseudomomentum (2.48) are *canonical equations* that involve *all* fields and degrees of freedom simultaneously. If this is obvious for the energy equation (2.18), which will contain all other fields in a more complex coupled-field theory, the situation is less clear for the equation of pseudomomentum, which is, after all, an equation of *linear momentum*. But the canonical nature of this is a result of the fact that it is generated by - or dual to - translations in material space. That is, Eq. (2.48) reflects the invariance or lack of invariance of the *whole* physical system under material-space translations. Obviously then, the balance of energy and pseudomomentum are indeed time and space-like conservation equations associated with the $(t, \mathbf{X})$ parametrization of the "material" mechanics of continua.

A final remark is in order concerning the expressions of $\mathcal{H}$ and $\mathcal{L}$. Although no variational principle was introduced, these two quantities are akin to Hamiltonian and Lagrangian densities, respectively, in analytical continuum mechanics. The relationship between the two therefore involves not only the usual Legendre transformation concerning kinetics but also the Legendre transformation germane to thermal processes, so that in all we have the double Legendre transformation [Maugin and Berezovski (1999)]

$$\mathcal{H} = \mathcal{P} \cdot \mathbf{V} + S\theta - \mathcal{L}. \tag{2.49}$$

2.5   Brittle fracture

It is clear that at any *regular* material point $\mathbf{X}$ the local equations of momentum (2.17) and (2.48) are, through the Ericksen's identity, direct consequences of one another. However, if we integrate each of these over a regular (simply connected) *homogeneous* material body $\mathcal{B}$ – of boundary $\partial\mathcal{B}$ equipped with unit outward normal $\mathbf{N}$ – thanks to the trivial commuting of material integration and material time differentiation and the mathematically justified use of the divergence theorem, we obtain the following two global equations:

$$\frac{d}{dt} \int_{\mathcal{B}} \mathbf{p} \, dV = \int_{\partial \mathcal{B}} \mathbf{N} \cdot \mathbf{T} \, dA, \qquad (2.50)$$

and

$$\frac{d}{dt} \int_{\mathcal{B}} \mathcal{P} \, dV = \int_{\partial \mathcal{B}} \mathbf{N} \cdot \mathbf{b} \, dA + \int_{\mathcal{B}} (\mathbf{f}^{inh} + \mathbf{f}^{th}) \, dV, \qquad (2.51)$$

respectively, where the second must be understood *component wise* (on the material manifold) because, in contrast to what happens in Eq. (2.50), the quantities involved in the integrands in Eq. (2.51) are material *covectors* and the material manifold $\mathcal{M}^3$ does not play the same neutral role as that played by the Euclidean physical manifold E^3 in Eq. (2.50). Obviously, in writing the global forms, we lost the convection property relating the two equations of linear momentum. This means that Eqs. (2.50) and (2.51) can be used for different purposes, essentially Eq. (2.50) for *solving* the physical initial-boundary value problem and Eq. (2.51) for the construction of *forces acting on defects* and devising a *criterion* for the progress of such defects. Equation (2.51) must also be compatible with the global energy conservation equation (2.11) for the relevant material region

$$\frac{d}{dt} \int_{\mathcal{B}} \mathcal{H} \, dV = \int_{\partial \mathcal{B}} \mathbf{N} \cdot (\mathbf{T} \cdot \mathbf{v} - \mathbf{Q}) \, dA. \qquad (2.52)$$

In a field-theoretical vision, defects correspond to *mathematical singularities of fields* in a continuous description, essentially the stress and strain fields in elasticity. An important subject of contemporary theoretical and applied mechanics is that of *fracture*. The most relevant notions developed by engineers in that context are those of a *driving force acting on the tip of a crack* and of *energy-release rate*. These two receive their best framework in the thermomechanics corresponding to the exploitation of global equations of the types (2.51) and (2.52). The reason for this is that Eq. (2.50), although global, does not capture the singularity of the mechanical field at the tip of the crack. But Eqs. (2.51) and (2.52) do the job because they involve integrands which have a higher degree of singularity. Furthermore, the notion of a driving force acting on the tip of a crack (a set of zero measure) can only be defined by a limiting procedure applied to a more global quantity, for example, an integral over a material region surrounding the singularity. This justifies the use of Eqs. (2.51) and (2.52), *jointly* and not only one of them as commonly thought in these considerations.

2.5.1   *Straight brittle crack*

To illustrate our viewpoint, consider that the body $\mathcal{B}$ presents a straight through notch made of two flat faces $\mathcal{S}^{\pm}$ distant of $2d$, parallel to the X_1-axis, and ending in the body in a semicylinder $\Gamma^*(d)$ of radius d (Fig. 2.2). The body is infinite along the cylinder generators, so the problem is essen-

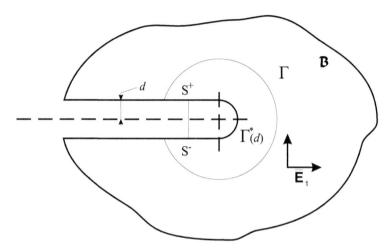

Fig. 2.2   Notch in an elastic body.

tially two-dimensional. The front of the notch is supposed to progress as a rigid surface, keeping the two flat faces parallel and free of loading, with a material velocity $\bar{\mathbf{V}}(\mathbf{X}, t)$. That is, we have the following *traction-free, adiabatic* conditions

$$\mathbf{N} \cdot \mathbf{T} = 0 \quad \text{and} \quad \mathbf{N} \cdot \mathbf{Q} = 0 \quad \text{at} \quad \mathcal{S}^{\pm} \cup \Gamma^*(d). \tag{2.53}$$

The material body $\mathcal{B}$ of regular boundary $\partial \mathcal{B} \cup \Gamma^*(d)$ - apart from the flat faces that are not acted upon and such that $\bar{\mathbf{V}} \cdot \mathbf{N} = 0$ and, therefore, play no role - is simply connected and present no field singularities inside it, so that global formulas of the types Eqs. (2.51) and (2.52) hold good except for the fact that one part of the boundary surface is moving and the generalized Reynolds transport theorem needs to be implemented. The results of the integration Eqs. (2.51) and (2.52) over this "notched" body are

$$\frac{d}{dt} \int_{\mathcal{B}} \mathcal{P}_I \, dV + F_I = \int_{\partial \mathcal{B}} (\mathbf{N} \cdot \mathbf{b})_I \, dA + \int_{\mathcal{B}} (\mathbf{f}^{inh} + \mathbf{f}^{th})_I \, dV, \tag{2.54}$$

and

$$\frac{d}{dt} \int_{\mathcal{B}} \mathcal{H} \, dV + G = \int_{\partial \mathcal{B}} \mathbf{N} \cdot (\mathbf{T} \cdot \mathbf{v} - \mathbf{Q}) \, dA, \qquad (2.55)$$

where we were cautious to write the first in component (I) form and observe the appearance of source terms due to the inward motion of the crack front. These are given by

$$F_I = \int_{\Gamma^*} (-\mathcal{L} N_I + \mathcal{P}_I (\bar{\mathbf{V}} \cdot \mathbf{N})) \, dA, \qquad (2.56)$$

and

$$G = \int_{\Gamma^*} \mathcal{H} (\bar{\mathbf{V}} \cdot \mathbf{N}) \, dA, \qquad (2.57)$$

on account of Eq. (2.53) [Dascalu and Maugin (1995)]. These are the components of a material force and an energy-release rate, respectively. Notice that the normal $\mathbf{N}$ is oriented *inward* along Γ^*.

Another way to compute F_I and G is to integrate Eqs. (2.51) and (2.52) over a subbody $\mathcal{C}$ of $\mathcal{B}$ that is bounded by a surface Γ around and moving with the crack tip, ending on both straight faces $\mathcal{S}^{\pm}$. Accounting for expressions (2.54) and (2.55) we then obtain the global balance of pseudomomentum and energy over $\mathcal{C}$ as

$$F_I = \int_{\Gamma} (\mathbf{N} \cdot \mathbf{b})_I \, dA + \int_{\mathcal{C}} (\mathbf{f}^{inh} + \mathbf{f}^{th})_I \, dV - \frac{d}{dt} \int_{\mathcal{C}} \mathcal{P}_I \, dV, \qquad (2.58)$$

and

$$G = \int_{\Gamma} \mathbf{N} \cdot (\mathbf{T} \cdot \mathbf{v} - \mathbf{Q}) \, dA - \frac{d}{dt} \int_{\mathcal{C}} \mathcal{H} \, dV. \qquad (2.59)$$

We recognize in the last expression minus the change in *potential energy (dynamic thermoelastic, homogeneous case)* changed of sign, that is, the *dissipation rate* in the progress of the notch inside the material body. Accordingly, there must be a relationship between the latter quantity and the *power expended* irreversibly *by the material force* of component F_I. Following Dascalu and Maugin (1993), we establish this relationship in the limit as the notch reduces to a straight through crack of zero thickness. In this limit, which is assumed to be uniform, we thus have the following reduced expressions for the sharp crack case as $d \to 0$

$$F_I = \lim_{d \to 0} \int_{\Gamma^*(d)} (-\mathcal{L} N_I + \mathcal{P}_I (\bar{\mathbf{V}} \cdot \mathbf{N})) \, dA, \qquad (2.60)$$

and

$$G = \lim_{d \to 0} \int_{\Gamma^*(d)} \mathcal{H}(\bar{\mathbf{V}} \cdot \mathbf{N}) \, dA. \tag{2.61}$$

Let $\mathbf{P}$ be a point of Γ^* and $\mathbf{X_p}$ its material coordinates. Let $\mathbf{r}$ be the position of a point with respect to $\mathbf{P}$. The general motion is written (compare to Eq. (2.1))

$$\chi(\mathbf{X}, t) = \chi(\mathbf{X}_p + \mathbf{r}, t) = \bar{\chi}(\mathbf{r}, t). \tag{2.62}$$

By differentiation with respect to time and convection back by $\mathbf{F}^{-1}$, this yields

$$\mathbf{V} = \bar{\mathbf{V}} - \mathbf{F}^{-1} \cdot \frac{\partial \bar{\chi}}{\partial t}, \quad \bar{\mathbf{V}} = V_I \mathbf{E}_I, \tag{2.63}$$

where $\mathbf{E}_I$ is a material basis. We obviously suppose all points of Γ^* are in uniform motion, that $\bar{\mathbf{V}} = \bar{\mathbf{V}}(t)$ only ("*en bloc*" motion as if Γ^* were rigid), so the second contribution in Eq. (2.63)$_1$ in fact vanishes. On account of this we can multiply Eq. (2.60) by $\bar{\mathbf{V}}$ and commute with the limit and insert in the integration to obtain

$$\bar{\mathbf{V}} \cdot F = \lim_{d \to 0} \int_{\Gamma^*(d)} (-\mathcal{L} + (\mathcal{P} \cdot \mathbf{V}))(\bar{\mathbf{V}} \cdot \mathbf{N}) \, dA. \tag{2.64}$$

Therefore, comparing to Eq. (2.61), we have

$$\bar{\mathbf{V}} \cdot F = G - \lim_{d \to 0} \int_{\Gamma^*(d)} (\theta S)(\bar{\mathbf{V}} \cdot \mathbf{N}) \, dA, \tag{2.65}$$

on account Eq. (2.49). But the last quantity in Eq. (2.65) converges toward zero [Maugin (1992)], so

$$\bar{\mathbf{V}} \cdot F = G \geq 0, \tag{2.66}$$

where the inequality sign indicates the thermodynamical irreversibility of the crack growth phenomenon. This corresponds to the presence of a *hot heat source* at the crack tip, an effect that can be observed via infrared thermography [Maugin (1992)].

The perspicacious reader will have noticed that expression (2.58) – the component along the crack direction – is a dynamical thermoelastic generalization of the celebrated *Eshelby-Cherepanov-Rice J*-integral of brittle fracture (e.g., [Rice (1968)]) if we remember the expression of the dynamical Eshelby tensor $\mathbf{b}$. The present formulation is strictly based on thermomechanical arguments on the material manifold.

2.6 Phase-transition fronts

In the case of phase transition fronts we consider the possible irreversible progress of a phase into another one, the separation between the two phases being idealized as a sharp, discontinuity surface $\mathcal{S}(t)$ across which most of the fields suffer finite discontinuity jumps. A phase transition consists of a local *material rearrangement* of the material and, therefore, a "material" description based on the exploitation of the Eshelby stress imposes itself. The presence of $\mathcal{S}(t)$ brakes the material symmetry of the material body as a whole and manifests a *material inhomogeneity*. Consequently, the critical equation for the description of the phenomenon is not only that relating to dissipation but also the equation associated with the *lack of conservation* of pseudomomentum across $\mathcal{S}(t)$, that is, the associated *jump relation* that is generated by a change of "particle" on the material manifold: the driving force acting on the transition front will, therefore, be a "material force".

2.6.1 *Jump relations*

The main problem here consists of writing down a consistent set of thermo-mechanical jump relations across the front $\mathcal{S}(t)$. The phase transition fronts considered are homothermal (no jump in temperature; the two phases co-exist at the same temperature) and coherent (they present no defects such as dislocations). Consequently, we have the following continuity conditions [Maugin and Trimarco (1995); Maugin (1997, 1998)]:

$$[\theta] = 0 \quad \text{and} \quad [\mathbf{V}] = 0 \quad \text{at} \quad \mathcal{S}. \tag{2.67}$$

Jump relations associated with strict conservation laws (2.23)-(2.25)

$$\bar{V}_N[\rho_0] = 0, \tag{2.68}$$

$$\bar{V}_N[\mathbf{p}] + \mathbf{N} \cdot [\mathbf{T}] = 0, \tag{2.69}$$

$$\bar{V}_N[\mathcal{H}] + \mathbf{N} \cdot [\mathbf{T} \cdot \mathbf{v} - \mathbf{Q}] = 0, \tag{2.70}$$

are complemented by jump relations for entropy and pseudomomentum with formally added unknown source terms to the strict conservative form [Maugin (1997, 1998)]

$$\bar{V}_N[S] - \mathbf{N} \cdot [\mathbf{Q}/\theta] = \sigma_\mathcal{S} \geq 0, \tag{2.71}$$

$$\bar{V}_N[\mathcal{P}] + \mathbf{N} \cdot [\mathbf{b}] = -\mathbf{f}_S. \qquad (2.72)$$

Accordingly, the interesting relationship is the one that relates the *unknown* driving force $\mathbf{f}_S$ and the equally unknown (but non-negative) surface entropy source σ_S. This last condition is the requirement of the second law of thermodynamics. If the theory is *consistent*, these cannot be entirely independent. The looked for consistency condition in fact allows one to close the system of phenomenological equations at S in compliance with the second law.

2.6.2 *Driving force*

To proceed toward this aim, we evaluate the power expended by the material surface force of inhomogeneity $\mathbf{f}_S$ in a motion of $S(t)$ with the velocity $\bar{\mathbf{V}}_S$

$$P_S := \mathbf{f}_S \cdot \bar{\mathbf{V}}_S. \qquad (2.73)$$

According to the jump relation for the pseudomomentum (2.72), we have

$$\mathbf{f}_S \cdot \bar{\mathbf{V}}_S = -\bar{V}_N[\mathcal{P}] \cdot \bar{\mathbf{V}}_S - \mathbf{N} \cdot [\mathbf{b}] \cdot \bar{\mathbf{V}}_S. \qquad (2.74)$$

Accounting for the continuity condition $(2.67)_2$ one can write

$$\mathbf{f}_S \cdot \bar{\mathbf{V}}_S = -[\bar{V}_N \mathcal{P} \cdot \mathbf{V}] - [\mathbf{N} \cdot (\mathbf{b} \cdot \mathbf{V})]. \qquad (2.75)$$

Remembering the definition of the pseudomomentum we can calculate

$$\mathcal{P} \cdot \mathbf{V} = -(\rho_0 \mathbf{v} \cdot \mathbf{F}) \cdot \mathbf{V} = \rho_0 \mathbf{v}^2. \qquad (2.76)$$

Further, according to the definition of the Eshelby stress tensor

$$\mathbf{b} := -(\frac{1}{2}\rho_0 \mathbf{v}^2 - W)\mathbf{I} - \mathbf{T} \cdot \mathbf{F}, \qquad (2.77)$$

we arrive at

$$\mathbf{b} \cdot \mathbf{V} = -(\frac{1}{2}\rho_0 \mathbf{v}^2 - W)\mathbf{V} - (\mathbf{T} \cdot \mathbf{F}) \cdot \mathbf{V} = -(\frac{1}{2}\rho_0 \mathbf{v}^2 - W)\mathbf{V} + \mathbf{T} \cdot \mathbf{v}. \quad (2.78)$$

Substituting Eqs. (2.76) and (2.78) into Eq. (2.75) we have then

$$\mathbf{f}_S \cdot \bar{\mathbf{V}}_S = -[\bar{V}_N \rho_0 \mathbf{v}^2] + [\mathbf{N} \cdot (\frac{1}{2}\rho_0 \mathbf{v}^2 - W)\mathbf{V} - \mathbf{N} \cdot (\mathbf{T} \cdot \mathbf{v})], \qquad (2.79)$$

or

$$P_S = \mathbf{f}_S \cdot \bar{\mathbf{V}}_S = -[\bar{V}_N(\frac{1}{2}\rho_0 \mathbf{v}^2 + W) + \mathbf{N} \cdot (\mathbf{T} \cdot \mathbf{v})]. \qquad (2.80)$$

On the other hand, we extract the scalar quantity $\mathbf{N} \cdot [\mathbf{Q}]$ from the jump relation for energy (2.70)

$$\mathbf{N} \cdot [\mathbf{Q}] = \bar{V}_N [\mathcal{H}] + \mathbf{N} \cdot [\mathbf{T} \cdot \mathbf{v}]. \tag{2.81}$$

By means of the homothermality condition $(2.67)_1$, we can represent the jump relation for entropy (2.71) in the form

$$\bar{V}_N [\theta S] - \mathbf{N} \cdot [\mathbf{Q}] = \theta \sigma_{\mathcal{S}}. \tag{2.82}$$

Substituting then $\mathbf{N} \cdot [\mathbf{Q}]$ from Eq. (2.82) into Eq. (2.81) we obtain

$$\bar{V}_N [\theta S] - \bar{V}_N [\mathcal{H}] - \mathbf{N} \cdot [\mathbf{T} \cdot \mathbf{v}] = \theta \sigma_{\mathcal{S}}. \tag{2.83}$$

Remembering the definition of the free energy density

$$W = E - \theta S, \tag{2.84}$$

we thus have

$$-[\bar{V}_N (\frac{1}{2} \rho_0 \mathbf{v}^2 + W) + \mathbf{N} \cdot (\mathbf{T} \cdot \mathbf{v})] = \theta \sigma_{\mathcal{S}}. \tag{2.85}$$

Comparing Eqs. (2.80) and (2.85) we conclude

$$P_{\mathcal{S}} = \mathbf{f}_{\mathcal{S}} \cdot \bar{\mathbf{V}}_{\mathcal{S}} = \theta \sigma_{\mathcal{S}}. \tag{2.86}$$

Now we complete the computation by evaluating $P_{\mathcal{S}}$ further as follows. We have

$$[\mathbf{N} \cdot (\mathbf{T} \cdot \mathbf{v})] = [\mathbf{N} \cdot \mathbf{T}] \cdot \langle \mathbf{v} \rangle + \langle \mathbf{N} \cdot \mathbf{T} \rangle \cdot [\mathbf{v}], \tag{2.87}$$

while from the jump relation for linear momentum (2.69)

$$\mathbf{N} \cdot [\mathbf{T}] \cdot \langle \mathbf{v} \rangle = -\bar{V}_N [\rho_0 \mathbf{v}] \cdot \langle \mathbf{v} \rangle = -\bar{V}_N \left[\frac{1}{2} \rho_0 \mathbf{v}^2 \right]. \tag{2.88}$$

Therefore, Eq. (2.87) can be rewritten as follows

$$[\mathbf{N} \cdot (\mathbf{T} \cdot \mathbf{v})] = -\bar{V}_N \left[\frac{1}{2} \rho_0 \mathbf{v}^2 \right] + \langle \mathbf{N} \cdot \mathbf{T} \rangle \cdot [\mathbf{v}], \tag{2.89}$$

or, equivalently,

$$[\mathbf{N} \cdot (\mathbf{T} \cdot \mathbf{v})] = -\bar{V}_N \left[\frac{1}{2} \rho_0 \mathbf{v}^2 \right] - \langle \mathbf{N} \cdot \mathbf{T} \rangle \cdot [\mathbf{F} \cdot \mathbf{N}] \bar{V}_N. \tag{2.90}$$

The last term follows from the application of the Maxwell-Hadamard lemma [Maugin and Trimarco (1995)]

$$[\mathbf{v}] = -[\mathbf{F} \cdot \mathbf{N}]\bar{V}_N. \tag{2.91}$$

Gathering the partial results (2.85), (2.86), and (2.82), we obtain

$$P_S = \mathbf{f}_S \cdot \bar{\mathbf{V}}_S = -\bar{V}_N[W - \langle \mathbf{N} \cdot \mathbf{T} \rangle \cdot (\mathbf{F} \cdot \mathbf{N})]. \tag{2.92}$$

The same can be expressed as

$$P_S = \mathbf{f}_S \cdot \bar{\mathbf{V}}_S = \bar{V}_N f_S, \tag{2.93}$$

where the scalar driving force f_S is introduced

$$f_S = -[W - \langle \mathbf{N} \cdot \mathbf{T} \rangle \cdot (\mathbf{F} \cdot \mathbf{N})]. \tag{2.94}$$

The result (2.94) is universal in so far as the continuity conditions (2.67) are fulfilled; no hypothesis of small strain or quasi-statics has been envisaged [Maugin (2000)].

2.7 On the exploitation of Eshelby's stress in isothermal and adiabatic conditions

The Eshelby stress tensor $\mathbf{b}$ is defined above on the basis of the free energy per unit reference volume, $W(\mathbf{F}, \theta)$:

$$\mathbf{b} = -(\mathcal{L}\mathbf{I} + \mathbf{T} \cdot \mathbf{F}), \tag{2.95}$$

wherein

$$\mathbf{T} = \left(\frac{\partial W}{\partial \mathbf{F}}\right)_\theta, \qquad S = -\left(\frac{\partial W}{\partial \theta}\right)_{\mathbf{F}}, \tag{2.96}$$

$\mathbf{F}$ is the deformation gradient, $\mathbf{T}$ is the first Piola-Kirchhoff stress tensor, $\mathcal{L} = K - W$ is the Lagrangian density of energy per unit reference volume, K is the kinetic energy density, and S is the entropy density per unit reference volume.

Here (in thermoelasticity), $\mathbf{b}$ satisfies the equation of balance of pseudo-momentum (we discard material inhomogeneities and external body forces for the sake of simplicity; cf. Eq. (2.48)):

$$\left.\frac{\partial \mathcal{P}}{\partial t}\right|_{\mathbf{X}} - div_R \mathbf{b} = \mathbf{f}^{th}, \tag{2.97}$$

where

$$\mathcal{P} = -\rho_0 \mathbf{v} \cdot \mathbf{F}, \qquad \mathbf{f}^{th} = S \nabla_R \theta. \tag{2.98}$$

To be *more precise in the notation* we should write $\mathbf{T}_W$ and $\mathbf{b}_W$ instead of $\mathbf{T}$ and $\mathbf{b}$ since the latter are built on W. Thus Eqs. $(2.96)_1$, (2.97) and $(2.98)_2$ should be written as

$$\mathbf{T}_W = \left(\frac{\partial W}{\partial \mathbf{F}}\right)_\theta, \quad \left.\frac{\partial \mathcal{P}}{\partial t}\right|_\mathbf{X} - div_R \mathbf{b}_W = \mathbf{f}_W^{th}, \quad \mathbf{f}_W^{th} = S\nabla_R \theta. \quad (2.99)$$

The above consideration of W is well adapted to the cases when homothermal conditions prevail. As is well known, for adiabatic conditions, one better considers as reference energy the *internal energy density* E, related to W by the Legendre transformation

$$W = E - \theta S, \quad (2.100)$$

where temperature

$$\theta = \left(\frac{\partial E}{\partial S}\right)_\mathbf{F}, \quad (2.101)$$

must be non-negative according to the physical meaning of the thermodynamical temperature. Hence $E(S, \mathbf{F}_{fixed})$ must be an increasing function of S. Obviously, we can define the Piola-Kirchhoff stress tensor in terms of E, being careful to use the following notation

$$\mathbf{T}_E = \left(\frac{\partial E}{\partial \mathbf{F}}\right)_S. \quad (2.102)$$

It follows from Eqs. $(2.99)_1$, (2.100), and (2.102) that

$$\mathbf{T}_W = \left(\frac{\partial W}{\partial \mathbf{F}}\right)_\theta = \left(\frac{\partial E}{\partial \mathbf{F}}\right)_\theta + \theta \left(\frac{\partial S}{\partial \mathbf{F}}\right)_\theta = \left(\frac{\partial E}{\partial \mathbf{F}}\right)_S = \mathbf{T}_E = \mathbf{T}. \quad (2.103)$$

Substituting both Eqs. (2.100) and (2.103) into Eq. (2.95), we have

$$\mathbf{b}_W = -((K - E + \theta S)\mathbf{I} + \mathbf{T}_E \cdot \mathbf{F}), \quad (2.104)$$

or, equivalently,

$$\mathbf{b}_W = \mathbf{b}_E - (\theta S)\mathbf{I}, \quad (2.105)$$

if the Eshelby stress tensor based on E is obviously defined as

$$\mathbf{b}_E = -((K - E)\mathbf{I} + \mathbf{T}_E \cdot \mathbf{F}). \quad (2.106)$$

Defining a thermal material force $\mathbf{f}_E^{th}$ by

$$\mathbf{f}_E^{th} = -\theta \nabla_R S, \qquad (2.107)$$

we obtain that the material momentum equation $(2.99)_2$ takes on the following form

$$\left.\frac{\partial \mathcal{P}}{\partial t}\right|_{\mathbf{X}} - div_R \mathbf{b}_E = \mathbf{f}_E^{th}, \qquad (2.108)$$

since

$$\mathbf{f}_W^{th} = div_R(S\theta\mathbf{I}) + \mathbf{f}_E^{th}. \qquad (2.109)$$

Equation (2.108) is that equation to be used in almost adiabatic conditions, when we know the adiabatic elasticity coefficients. Strictly in uniform adiabatic conditions, $\mathbf{f}_E^{th} \equiv \mathbf{0}$.

2.7.1 *Driving force at singular surface in adiabatic conditions*

In order to write the jump relation associated with the balance of material momentum at a singular surface S, we simply used the formalism of hyperbolic systems for the terms in divergence and time derivative and added an as yet unknown surface material force to be determined. That is, together with the balance equation $(2.99)_2$ at regular material points in the body, at S we had the following jump relation:

$$\bar{V}_N[\mathcal{P}] + \mathbf{N} \cdot [\mathbf{b}_W] = -\mathbf{f}_{S,W}. \qquad (2.110)$$

The unknown driving force present in that equation, in the spirit of d'Alembert's principle, is evaluated by computing the dissipated power $\mathbf{f}_{S,W} \cdot \bar{\mathbf{V}}$ that it expends in its motion of material velocity $\bar{\mathbf{V}}$.

In the adiabatic case, we could as well have proposed the jump relation

$$\bar{V}_N[\mathcal{P}] + \mathbf{N} \cdot [\mathbf{b}_E] = -\mathbf{f}_{S,E} \qquad (2.111)$$

at S, where the "source" term is a priori different from the one in Eq. (2.110). On substituting from Eq. (2.105) into Eq. (2.110), we find that

$$\bar{V}_N[\mathcal{P}] + \mathbf{N} \cdot [\mathbf{b}_E] = -\mathbf{f}_{S,W} + \mathbf{N}[S\theta], \qquad (2.112)$$

from which we deduce that

$$\mathbf{f}_{S,E} = \mathbf{f}_{S,W} - \mathbf{N}[S\theta], \qquad (2.113)$$

an expression which is in harmony with Eqs. (2.108)-(2.109) if we replace formally the divergence by the normal jump.

2.7.2   *Another approach to the driving force*

Evaluating the power expended by the material surface force of inhomogeneity $\mathbf{f}_S$ in a motion of $S(t)$ with the velocity $\bar{\mathbf{V}}_S$ by means of the jump relation for the pseudomomentum (2.110), we have

$$P_S = \mathbf{f}_{S,W} \cdot \mathbf{V}_S = -[\bar{V}_N(\frac{1}{2}\rho_0\mathbf{v}^2 + W) + \mathbf{N} \cdot (\mathbf{T} \cdot \mathbf{v})]. \qquad (2.114)$$

On the other hand, we extract the scalar quantity $\mathbf{N} \cdot [\mathbf{Q}]$ from the jump relation for energy (2.70)

$$\mathbf{N} \cdot [\mathbf{Q}] = \bar{V}_N [\mathcal{H}] + \mathbf{N} \cdot [\mathbf{T} \cdot \mathbf{v}]. \qquad (2.115)$$

We can represent the jump relation for entropy (2.71) in the form

$$\bar{V}_N[S] - \mathbf{N} \cdot [\mathbf{Q}] \left\langle \frac{1}{\theta} \right\rangle - \mathbf{N} \cdot \langle \mathbf{Q} \rangle \left[\frac{1}{\theta} \right] = \sigma_S. \qquad (2.116)$$

As it was noted by Abeyaratne and Knowles (2000), both in the adiabatic and non-adiabatic cases

$$\mathbf{Q}[\theta] = \mathbf{0} \qquad (2.117)$$

at $S(t)$ (this means that either one of the two factors is zero separately and thus defines the two situations), and therefore necessarily

$$\mathbf{N} \cdot [\mathbf{Q}] \left(\left\langle \frac{1}{\theta} \right\rangle - \frac{1}{\langle\theta\rangle} \right) = 0 \quad \text{and} \quad \mathbf{N} \cdot \langle \mathbf{Q} \rangle \left[\frac{1}{\theta} \right] = 0. \qquad (2.118)$$

Substituting $\mathbf{N} \cdot [\mathbf{Q}]$ from Eq. (2.81) into Eq. (2.82) and taking into account Eqs. (2.117) and (2.118) we obtain

$$\bar{V}_N \langle\theta\rangle[S] - \bar{V}_N [\mathcal{H}] - \mathbf{N} \cdot [\mathbf{T} \cdot \mathbf{v}] = \langle\theta\rangle\sigma_S. \qquad (2.119)$$

Remembering the definition of the free energy density (2.100) we have

$$\bar{V}_N \langle\theta\rangle[S] - \bar{V}_N [\theta S] - [\bar{V}_N(\frac{1}{2}\rho_0\mathbf{v}^2 + W) + \mathbf{N} \cdot (\mathbf{T} \cdot \mathbf{v})] = \langle\theta\rangle\sigma_S. \qquad (2.120)$$

Comparing Eqs. (2.80) and (2.85) we conclude

$$\mathbf{f}_{S,W} \cdot \bar{\mathbf{V}}_S - \bar{V}_N \langle S \rangle [\theta] = \langle\theta\rangle\sigma_S. \qquad (2.121)$$

Since $\bar{V}_N = \mathbf{N} \cdot \bar{\mathbf{V}}_S$, the latter relation can be rewritten as

$$(\mathbf{f}_{S,W} - \mathbf{N}\langle S \rangle [\theta]) \cdot \bar{\mathbf{V}}_S = \langle\theta\rangle\sigma_S. \qquad (2.122)$$

The expression within parentheses on the left-hand side of the latter equation is what is considered as driving force by Abeyaratne and Knowles (2000). In isothermal processes it obviously coincides with $\mathbf{f}_{S,W}$, and in adiabatic processes [Abeyaratne and Knowles (2000)]

$$\mathbf{f}_{S,W} - \mathbf{N}\langle S \rangle [\theta] = \mathbf{N}\langle \theta \rangle [S], \qquad (2.123)$$

which leads to

$$\mathbf{f}_{S,W} = \mathbf{N}[S\theta], \qquad (2.124)$$

in full correspondence with Eq. (2.113) since $\mathbf{f}_{S,E} = \mathbf{0}$ in adiabatic processes.

2.8   Concluding remarks

The canonical formulation of the thermomechanics of elastic conductors of heat on the material manifold clearly plays a fundamental role, not in solving the original field problem, but in evaluating thermodynamically admissible driving forces, such as in the cases of brittle fracture and the propagation of phase transition fronts. These fictitious (following the remark of Eshelby) forces place in evidence the quasi-inhomogeneity in the continuum mechanics expressed directly on the material manifold, that is, where the geometrization of such mechanics is most effective. Furthermore, the canonical formulation thus emphasized links inevitably mechanics and dissipative effects of a topological nature as shown by the establishment of compatibility conditions between the expression of material driving forces and global dissipation. Finally, the presence of additional balance laws that would be redundant in many cases (in the absence of field singularities) provides checking means of numerical schemes or, even better, the framework for an original numerical scheme.

Chapter 3

Local Phase Equilibrium and Jump Relations at Moving Discontinuities

The material formulation of continuum mechanics gives us the general basis for a description of the dynamic behavior of inhomogeneous media including discontinuities. The governing equations are balance laws with source terms and jump relations at discontinuities. However, in the case of moving discontinuities, such as crack fronts and phase-transition boundaries, the jump relations depend on the velocity of the discontinuity, which remains undetermined.

In the case of phase transformations, this situation is well described by Anderson et al. (2007): "When a phase transformation does occur, there is an additional kinematical degree of freedom represented by the motion of the interface relative to the material; because of this the interfacial expressions for balance of mass, momentum, and energy fail to provide a closed description: an additional interface condition is needed to account for the microphysics associated with the exchange of material between phases."

Contemporary attempts to improve and generalize the jump conditions at the discontinuities [Gurtin and Voorhees (1996); O'Reilly and Varadi (1999); Cermelli and Sellers (2000); Gurtin and Jabbour (2002); Irschik (2003); Fischer and Simha (2004)] left the velocity of the discontinuity undetermined. This indeterminacy relates to entropy production due to irreversible motion of the discontinuity. From the thermodynamical point of view, the presence of entropy production requires to revise the jump relations at the interface between two systems in non-equilibrium.

The main idea is to use the so-called *local phase equilibrium conditions* at the phase boundary instead of the classical equilibrium conditions. The difference between the local phase equilibrium conditions and the classical equilibrium conditions is due to the total entropy, which is conserved in the classical full equilibrium, but this is not the case in the local phase equilibrium. Therefore, the local phase equilibrium conditions should have a form which does not contain the entropy explicitly. As we shall see, this is impossible for thermodynamic potentials represented as functions of natural sets of independent variables.

If we change the set of natural variables for internal energy to another one, we will not, in general, yield all the thermodynamic properties of the system. Moreover, the properties of convexity of the internal energy may not be preserved. While this is not desired in most cases, the loss of convexity is acceptable in the description of the behavior of a two-phase material [Ericksen (1998)]. Having looked at the intrinsic stability of simple thermodynamic systems in the forthcoming section, we shall be able to specify what is to be understood by "local phase equilibrium" in section 3.2, and then we can proceed with non-equilibrium states and what happens more particularly at a discontinuity surface.

3.1   Intrinsic stability of simple systems

We start with a reminder of the intrinsic stability conditions for single-component systems, which gives us some heuristic idea about possible sets of independent variables for internal energy in the case of local phase equilibrium.

A single-component fluid-like system is characterized by entropy S as the function of internal energy U, volume V and mass M of the system [Callen (1960)]

$$S = S(U, V, M). \tag{3.1}$$

The entropy maximum postulate yields that the hyper-surface $S = S(U, V, M)$ in the thermodynamic configuration space should have the property that it lies everywhere below its tangent planes. This implies conditions for the derivatives of the fundamental relation [Callen (1960)]

$$\left(\frac{\partial^2 S}{\partial U^2} \right)_{V,M} \leq 0, \tag{3.2}$$

$$\left(\frac{\partial^2 S}{\partial V^2} \right)_{U,M} \leq 0, \tag{3.3}$$

and

$$\frac{\partial^2 S}{\partial U^2} \frac{\partial^2 S}{\partial V^2} - \left(\frac{\partial^2 S}{\partial U \partial V} \right)^2 \geq 0. \tag{3.4}$$

The stability criteria can be reformulated in the energy representation $U = U(S, V, M)$. Whereas the entropy is maximum, the energy is minimum; thus the concavity of the entropy surface is replaced by convexity of energy surface. The local conditions of convexity become [Callen (1960)]

$$\left(\frac{\partial^2 U}{\partial S^2}\right)_{V,M} = \left(\frac{\partial T}{\partial S}\right)_{V,M} \geq 0, \tag{3.5}$$

$$\left(\frac{\partial^2 U}{\partial V^2}\right)_{S,M} = -\left(\frac{\partial p}{\partial V}\right)_{S,M} \geq 0. \tag{3.6}$$

It follows that the Helmholtz free energy $W = W(T, V, M)$ is a concave function of the temperature and convex function of the volume [Callen (1960)]

$$\left(\frac{\partial^2 W}{\partial T^2}\right)_{V,M} = -\left(\frac{\partial S}{\partial T}\right)_{V,M} \leq 0, \tag{3.7}$$

$$\left(\frac{\partial^2 W}{\partial V^2}\right)_{T,M} = -\left(\frac{\partial p}{\partial V}\right)_{T,M} \geq 0. \tag{3.8}$$

If the stability criteria are not satisfied, the system breaks into two or more phases. Therefore, it is supposed that the nucleation of a new phase in a single-component system corresponds to the marginal stability condition

$$\left(\frac{\partial^2 W}{\partial V^2}\right)_{T,M} = -\left(\frac{\partial p}{\partial V}\right)_{T,M} = 0. \tag{3.9}$$

This condition can be expressed in terms of the thermodynamic derivatives of the internal energy as follows

$$\left(\frac{\partial U}{\partial V}\right)_{T,M} = \left(\frac{\partial U}{\partial V}\right)_{p,M}. \tag{3.10}$$

In fact, comparing the total differentials of the energy

$$dU = \left(\frac{\partial U}{\partial V}\right)_{T,M} dV + \left(\frac{\partial U}{\partial T}\right)_{V,M} dT + \left(\frac{\partial U}{\partial M}\right)_{T,V} dM, \tag{3.11}$$

and

$$dU = \left(\frac{\partial U}{\partial V}\right)_{p,M} dV + \left(\frac{\partial U}{\partial p}\right)_{V,M} dp + \left(\frac{\partial U}{\partial M}\right)_{p,V} dM, \tag{3.12}$$

we have by account of Eq. (3.10)

$$\frac{dp}{dT} = \left(\frac{\partial p}{\partial T}\right)_{V,M},\qquad(3.13)$$

since the mass of the system remains unchanged. The latter means that the pressure is independent of the volume ($p = p(T, M)$), which gives Eq. (3.9).

Thus, the marginal stability condition can be expressed in terms of thermodynamic derivatives of the internal energy with respect to volume at fixed temperature and pressure, respectively.

3.2   Local phase equilibrium

After the nucleation of the new phase, the two-phase material consists of two single-component fluid-like systems separated by a sharp interface.

3.2.1   *Classical equilibrium conditions*

First we recall how the classical equilibrium conditions of single-component fluid-like systems are derived.

Consider two single-component fluid-like systems separated by a sharp interface and surrounded by the environment which is the same for both systems. In general the systems are in non-equilibrium whereas the environment is presupposed to be in equilibrium because of its reservoir properties. To be able to use common thermodynamic parameters, we suppose following [Kestin (1992)] that the non-equilibrium state of each system can be associated with a local equilibrium state of the accompanying thermostatic system.

In the case of a single-component system, we must know the values of three variables for the complete description of the equilibrium state of the system. In general, the states of the two systems may have different entropies $S^{(1)}$ and $S^{(2)}$, volumes $V^{(1)}$ and $V^{(2)}$, and masses $M^{(1)}$ and $M^{(2)}$

$$S^{(1)} \neq S^{(2)}, \quad V^{(1)} \neq V^{(2)}, \quad M^{(1)} \neq M^{(2)}.\qquad(3.14)$$

Consider also the composite system which combines both systems 1 and 2. Suppose that the composite system is in full equilibrium with the environment. Then

$$dU_{cs} = 0, \quad dS_{cs} = 0, \quad dV_{cs} = 0, \quad dM_{cs} = 0,\qquad(3.15)$$

where U denotes the internal energy and subscript "cs" denotes the composite system.

To derive the classical equilibrium conditions we consider the sum of total differentials of the internal energy for both systems

$$
\begin{aligned}
dU^{(i)} = & \left(\frac{\partial U^{(i)}}{\partial S} \right)_{V,M} d\,S^{(i)} + \\
& + \left(\frac{\partial U^{(i)}}{\partial V} \right)_{S,M} d\,V^{(i)} + \left(\frac{\partial U^{(i)}}{\partial M} \right)_{S,V} d\,M^{(i)}.
\end{aligned}
\tag{3.16}
$$

Due to additivity of extensive quantities

$$
\begin{aligned}
U_{cs} = U^{(1)} + U^{(2)}, \quad S_{cs} = S^{(1)} + S^{(2)}, \\
V_{cs} = V^{(1)} + V^{(2)}, \quad M_{cs} = M^{(1)} + M^{(2)},
\end{aligned}
\tag{3.17}
$$

and equilibrium conditions for the composite system (3.15), we obtain for this sum

$$
\begin{aligned}
& \left(\left(\frac{\partial U^{(1)}}{\partial S} \right)_{V,M} - \left(\frac{\partial U^{(2)}}{\partial S} \right)_{V,M} \right) d\,S^{(1)} \\
& + \left(\left(\frac{\partial U^{(1)}}{\partial V} \right)_{S,M} - \left(\frac{\partial U^{(2)}}{\partial V} \right)_{S,M} \right) d\,V^{(1)} \\
& + \left(\left(\frac{\partial U^{(1)}}{\partial M} \right)_{S,V} - \left(\frac{\partial U^{(2)}}{\partial M} \right)_{S,V} \right) d\,M^{(1)} = 0.
\end{aligned}
\tag{3.18}
$$

Since three independent variables can be varied arbitrary, we have zero jumps of the corresponding partial derivatives of the internal energy

$$
\left[\left(\frac{\partial U}{\partial S} \right)_{V,M} \right] = 0 \quad \text{or} \quad [\theta] = 0,
\tag{3.19}
$$

$$
\left[\left(\frac{\partial U}{\partial V} \right)_{S,M} \right] = 0 \quad \text{or} \quad [p] = 0,
\tag{3.20}
$$

$$
\left[\left(\frac{\partial U}{\partial M} \right)_{S,V} \right] = 0 \quad \text{or} \quad [\mu] = 0,
\tag{3.21}
$$

where θ is temperature, p is pressure and μ is chemical potential.

Thus, the classical equilibrium conditions consist, for fluid-like systems, of the equality of temperatures, pressures and chemical potentials in the two systems. The generalization of the equilibrium conditions to the case of solids can be found in Cermelli and Sellers (2000).

3.2.2   Local equilibrium jump relations

What we need are the conditions of local equilibrium between the two phases, since in the case of a moving phase boundary we never reach the full equilibrium state. Suppose that the local equilibrium states of the two phases are strongly different. This means that the values of thermodynamic state functions are not the same in the two systems

$$S^{(1)} \neq S^{(2)}, \quad V^{(1)} \neq V^{(2)}, \quad M^{(1)} \neq M^{(2)}. \tag{3.22}$$

It is clear that this difference in values of thermodynamic state functions is incompatible with classical equilibrium conditions (3.19)-(3.21). This incompatibility is expected, because our two systems are only in a local but not in the full equilibrium.

In the full equilibrium the total energy, the total entropy, and the total mass are conserved. However, we cannot expect the conservation of the total entropy in the local phase equilibrium, because the phase transformation process is accompanied with entropy production. Therefore, we need to derive the local phase equilibrium conditions in a form which does not contain the entropy explicitly. This is impossible for thermodynamic potentials represented as functions of natural sets of independent variables.

Possible sets of independent variables are either θ, V, M, or p, V, M, or θ, p, M. The third set is excluded, because there is no additivity for both temperatures and pressures. Therefore, we have only two possibilities.

Choosing the set of independent variables θ, V, M, we consider total differentials of the internal energy for both systems

$$
\begin{aligned}
dU^{(1)} = &\left(\frac{\partial U^{(1)}}{\partial \theta} \right)_{V,M} d\,\theta^{(1)} \\
&+ \left(\frac{\partial U^{(1)}}{\partial V} \right)_{\theta,M} d\,V^{(1)} + \left(\frac{\partial U^{(1)}}{\partial M} \right)_{\theta,V} d\,M^{(1)},
\end{aligned}
\tag{3.23}
$$

and

$$
\begin{aligned}
dU^{(2)} = &\left(\frac{\partial U^{(2)}}{\partial \theta} \right)_{V,M} d\,\theta^{(2)} \\
&+ \left(\frac{\partial U^{(2)}}{\partial V} \right)_{\theta,M} d\,V^{(2)} + \left(\frac{\partial U^{(2)}}{\partial M} \right)_{\theta,V} d\,M^{(2)}.
\end{aligned}
\tag{3.24}
$$

The conservation of mass and energy and the additivity of volume yields

$$dU^{(1)} = -dU^{(2)}, \quad d\,V^{(1)} = -d\,V^{(2)}, \quad d\,M^{(1)} = -d\,M^{(2)}. \tag{3.25}$$

Since masses and volumes can be varied arbitrarily, we then have

$$\left(\frac{\partial U^{(1)}}{\partial \theta}\right)_{V,M} d\theta^{(1)} + \left(\frac{\partial U^{(2)}}{\partial \theta}\right)_{V,M} d\theta^{(2)} = 0, \qquad (3.26)$$

$$\left[\left(\frac{\partial U}{\partial V}\right)_{\theta,M}\right] = 0, \qquad (3.27)$$

$$\left[\left(\frac{\partial U}{\partial M}\right)_{\theta,V}\right] = 0. \qquad (3.28)$$

The first of the obtained conditions means that any change in temperature of one of the subsystems should lead to a corresponding change in temperature in another subsystem in order to hold the local equilibrium.

With another choice of independent variables, namely, p, V, M, we have, correspondingly,

$$\left(\frac{\partial U^{(1)}}{\partial p}\right)_{V,M} dp^{(1)} + \left(\frac{\partial U^{(2)}}{\partial p}\right)_{V,M} dp^{(2)} = 0, \qquad (3.29)$$

$$\left[\left(\frac{\partial U}{\partial V}\right)_{p,M}\right] = 0, \qquad (3.30)$$

$$\left[\left(\frac{\partial U}{\partial M}\right)_{p,V}\right] = 0. \qquad (3.31)$$

Our special interest is in the conditions (3.27) and (3.30) since just these conditions correspond to mechanical equilibrium in the set of the classical relations.

The relation (3.27) can be represented in another form

$$\left[\left(\frac{\partial U}{\partial V}\right)_{\theta,M}\right] = \left[\theta\left(\frac{\partial p}{\partial \theta}\right)_{V,M} - p\right] = 0, \qquad (3.32)$$

and in the isothermal case it is reduced to the classical equilibrium condition (3.20), which is compatible with the Rankine-Hugoniot jump relations only for zero value of the velocity of discontinuity. This means that the local equilibrium condition (3.27) corresponds to a stationary phase boundary.

Another local equilibrium condition (3.30) can be rewritten in the form

$$\left[\left(\frac{\partial U}{\partial V}\right)_{p,M}\right] = \left[\theta\left(\frac{\partial S}{\partial V}\right)_{p,M} - p\right] = 0. \qquad (3.33)$$

This condition provides a non-zero jump in local pressures of distinct phases. The first term on the right-hand side of Eq. (3.33) can be interpreted as the energy release rate for the phase transformation. It is expected that just this local equilibrium condition is satisfied at the moving phase boundary.

It is remarkable that the thermodynamic derivatives in the local phase equilibrium conditions at the phase boundary are the same as those used in the condition of marginal stability (3.10).

At the point where the phase boundary starts to move, one can expect a continuous change of the set of conditions from one to the other. Therefore, the two sets should be simultaneously satisfied at this point.

3.3   Non-equilibrium states

The obtained local equilibrium jump relations are represented in terms of energy. In practice, however, we exploit more common quantities, like temperature, pressure, etc., which are related to the energy by constitutive equations. Therefore we need to have a description of non-equilibrium states.

It should be noted that there is no conventional description of non-equilibrium states, because even the notion of non-equilibrium temperature can be defined in different manners [Casas-Vázquez and Jou (2003)]. However, keeping in mind the possibility of numerical simulations, we choose the thermodynamics of discrete systems [Muschik (1993)] for the description of non-equilibrium states. In this theory, the state space of any system is extended by means of so-called *contact quantities*. In the simplest case of a fluid-like system, they are *contact temperature, Θ, dynamic pressure, π*, and *dynamic chemical potential, ν*. These quantities correspond to the interaction between a system and its surroundings by heat, work and mass exchanges, and determined by the following inequalities [Muschik (1993)]:

$$\dot{Q}\left(\frac{1}{\Theta} - \frac{1}{\theta^*}\right) \geq 0, \quad (\dot{V} = 0, \dot{M} = 0), \qquad (3.34)$$

$$\dot{V}\left(\pi - p^*\right) \geq 0, \quad (\dot{Q} = 0, \dot{M} = 0), \qquad (3.35)$$

$$\dot{M}\left(\mu^* - \nu\right) \geq 0, \quad (\dot{V} = 0, \dot{Q} = 0). \qquad (3.36)$$

Here θ^* is the thermostatic temperature of the equilibrium environment, p^* its equilibrium pressure, and μ^* its equilibrium chemical potential. The contact temperature is the dynamical analogue to the thermostatic temperature [Muschik (1993)]. The interpretation of the contact temperature is as follows: From Eq. (3.34) it is evident that the heat exchange $\dot{Q}$ and the bracket always have the same sign. We now presuppose that there exists exactly one equilibrium environment for each arbitrary state of a system for which the net heat exchange between them vanishes. Consequently the defining inequality (3.34) determines the contact temperature Θ of the system as that thermostatic temperature θ of the system's environment, for which this net heat exchange vanishes. The same interpretation holds true for the dynamical pressure π and the dynamical chemical potential ν with respect to the net rate of the volume and to the net rate of each mole number due to external material exchange.

Thus the complete description of the non-equilibrium state of a system includes both common thermodynamic parameters and contact quantities characterizing the interaction of the system with its environment.

3.4   Local equilibrium jump relations at discontinuity

The conventional procedure in the description of an irreversible process is a projection of a non-equilibrium state of a system onto a state of an accompanying local equilibrium system [Kestin (1992); Maugin (1999b)]. Theoretically, it is possible in principle to find a projection which provides identical values for local equilibrium parameters and for corresponding contact quantities of the non-equilibrium system. It is clear that we should point out which projection procedure is applied, before we use the local equilibrium quantities for the description of a non-equilibrium state.

In general, we need to take into account that the local equilibrium values may differ from the values of contact quantities [Muschik and Berezovski (2007)]. This difference results in the appearance of so-called *excess quantities* [Muschik and Berezovski (2004, 2007)]. For instance, in the required extension of the concepts of the thermodynamics of discrete systems to the *thermoelastic case* we expect for the entropy per unit volume

$$S = S_{eq} + S_{ex}. \tag{3.37}$$

Here S is the total non-equilibrium entropy, S_{eq} is its local equilibrium value, and S_{ex} is the excess of entropy.

The constitutive relations for thermoelastic conductors define the entropy in terms of free energy per unit volume

$$S = -\left(\frac{\partial W}{\partial \theta}\right)_{\mathbf{F}}, \tag{3.38}$$

taken at fixed deformation. We can define the entropy excess similarly

$$S_{ex} = -\left(\frac{\partial W_{ex}}{\partial \theta}\right)_{\mathbf{F}} \tag{3.39}$$

by introducing a free energy excess

$$W = W_{eq} + W_{ex}. \tag{3.40}$$

The latter leads immediately to the excess of the stress tensor because in thermoelasticity, by definition,

$$\mathbf{T} = \left(\frac{\partial W_{eq}}{\partial \mathbf{F}}\right)_{\theta}. \tag{3.41}$$

Consequently, we need to define the excess of the stress tensor $\mathbf{\Sigma}$ as a quantity associated with the excess of free energy

$$\mathbf{\Sigma} = \left(\frac{\partial W_{ex}}{\partial \mathbf{F}}\right)_{\theta}. \tag{3.42}$$

In the thermoelastic case, the thermodynamic derivatives which we should exploit instead of $\left(\frac{\partial U}{\partial V}\right)_{\theta}$ and $\left(\frac{\partial U}{\partial V}\right)_{p}$ in Eqs. (3.32) and (3.33) are obviously the following ones:

$$\left(\frac{\partial E_{eq}}{\partial \mathbf{F}}\right)_{\theta} = -\theta \left(\frac{\partial \mathbf{T}}{\partial \theta}\right)_{\mathbf{F}} + \mathbf{T}, \tag{3.43}$$

and

$$\left(\frac{\partial E_{eq}}{\partial \mathbf{F}}\right)_{\mathbf{T}} = \theta \left(\frac{\partial S_{eq}}{\partial \mathbf{F}}\right)_{\mathbf{T}} + \mathbf{T}, \tag{3.44}$$

where E is the internal energy per unit volume.

The introduced excess quantities are connected with the excess energy in a similar way

$$\left(\frac{\partial E_{ex}}{\partial \mathbf{F}}\right)_{\theta} = -\theta \left(\frac{\partial \mathbf{\Sigma}}{\partial \theta}\right)_{\mathbf{F}} + \mathbf{\Sigma}, \tag{3.45}$$

and

$$\left(\frac{\partial E_{ex}}{\partial \mathbf{F}}\right)_{\mathbf{T}} = \theta \left(\frac{\partial S_{ex}}{\partial \mathbf{F}}\right)_{\mathbf{T}} + \mathbf{\Sigma}. \tag{3.46}$$

Therefore, the local equilibrium jump relations (3.32) and (3.33) can be rewritten in terms of excess quantities (square brackets still denote jumps):

$$\left[-\theta \left(\frac{\partial \mathbf{T}}{\partial \theta} \right)_{\mathbf{F}} + \mathbf{T} - \theta \left(\frac{\partial \mathbf{\Sigma}}{\partial \theta} \right)_{\mathbf{F}} + \mathbf{\Sigma} \right] \cdot \mathbf{N} = 0, \qquad (3.47)$$

$$\left[\theta \left(\frac{\partial S_{eq}}{\partial \mathbf{F}} \right)_{\mathbf{T}} + \mathbf{T} + \theta \left(\frac{\partial S_{ex}}{\partial \mathbf{F}} \right)_{\mathbf{T}} + \mathbf{\Sigma} \right] \cdot \mathbf{N} = 0. \qquad (3.48)$$

As previously, the jump relation (3.47) should be applied in the case of stationary discontinuities, while the second jump relation (3.48) corresponds to moving discontinuities.

To make this new local equilibrium jump relations useful, we need to determine the jumps of the excess quantities at the discontinuity. Before the determination of the excess quantities at the singularity, we note that in the isothermal case without entropy production both of the local equilibrium jump relations are reduced to one simple relation

$$[\mathbf{T} + \mathbf{\Sigma}] \cdot \mathbf{N} = 0. \qquad (3.49)$$

In quasi-statics, the normal Cauchy traction should be zero

$$[\mathbf{T}] \cdot \mathbf{N} = 0, \qquad (3.50)$$

that means that the jump of the stress excess should be reduced in full correspondence to equilibrium state:

$$[\mathbf{\Sigma}] \cdot \mathbf{N} = 0. \qquad (3.51)$$

3.5   Excess quantities at a moving discontinuity

Due to the additivity of entropy, the local equilibrium jump relation (3.48) at a moving discontinuity can be rewritten as follows:

$$[\mathbf{\Sigma}] \cdot \mathbf{N} = -\left[\mathbf{T} + \theta \left(\frac{\partial S}{\partial \mathbf{F}} \right)_{\mathbf{T}} \right] \cdot \mathbf{N}. \qquad (3.52)$$

As shown by [Abeyaratne and Knowles (2000)], the rate of entropy production due to the propagating phase boundary in both adiabatic and nonadiabatic cases is determined only by the driving force and the velocity of the front. This means that the jump relation for the entropy (2.26) can be specialized as

$$\bar{V}_N [S] = \sigma_S \geq 0, \qquad (3.53)$$

where the entropy production at the discontinuity is given by [Maugin (1997, 1998)]

$$\theta_S \sigma_S = \mathbf{f} \cdot \mathbf{V} = f_S \bar{V}_N. \tag{3.54}$$

It follows from Eqs. (3.53), (3.54) that the jump of entropy is determined by the scalar value of the driving force and the temperature at the discontinuity

$$[S] = \frac{f_S}{\theta_S}. \tag{3.55}$$

To be able to calculate the derivative of the entropy in Eq. (3.52) we assume that in the vicinity of the discontinuity the entropy behaves like the driving force

$$S = \frac{f}{\theta}, \quad f_S = [f], \tag{3.56}$$

where the function f is defined similarly to Eq. (2.86)

$$f = -W + \mathbf{N} \cdot \langle \mathbf{T} \rangle \cdot \mathbf{F} \cdot \mathbf{N}. \tag{3.57}$$

This formally defined quantity – a generator function – acquires a physical meaning only when evaluated at a point $\mathbf{X}$ belonging to an oriented surface of unit normal $\mathbf{N}$ and with an operation such as that of taking the discontinuity applied to it.

Using the representation (3.56), we obtain for the derivative of the entropy with respect to the deformation gradient

$$\left(\frac{\partial S}{\partial \mathbf{F}} \right)_{\mathbf{T}} = \frac{1}{\theta} \left(\frac{\partial f}{\partial \mathbf{F}} \right)_{\mathbf{T}} - \frac{f}{\theta^2} \left(\frac{\partial \theta}{\partial \mathbf{F}} \right)_{\mathbf{T}}. \tag{3.58}$$

Substituting the last relation into Eq. (3.52), we can determine the jump of the stress excess at a discontinuity

$$[\mathbf{\Sigma}] \cdot \mathbf{N} = - [\mathbf{T}] \cdot \mathbf{N} + \left(f_S \left\langle \frac{1}{\theta} \left(\frac{\partial \theta}{\partial \mathbf{F}} \right)_{\mathbf{T}} \right\rangle \right. \\ \left. + \langle f \rangle \left[\frac{1}{\theta} \left(\frac{\partial \theta}{\partial \mathbf{F}} \right)_{\mathbf{T}} \right] - \left[\left(\frac{\partial f}{\partial \mathbf{F}} \right)_{\mathbf{T}} \right] \right) \cdot \mathbf{N}. \tag{3.59}$$

Though the jump of the stress excess at the discontinuity is determined, we still need certain additional assumption to fix the value of the velocity of the discontinuity. The simplest assumption is the continuity of the stress excess at the discontinuity (3.51)

$$[\mathbf{\Sigma}] \cdot \mathbf{N} = 0. \tag{3.60}$$

This condition removes the indeterminacy. If we keep the continuity of the stress excess across the discontinuity, we come to the relation between the jump of normal component of the Piola-Kirchhoff stress tensor and the driving force

$$
[\mathbf{T}] \cdot \mathbf{N} = \left(f_S \left\langle \frac{1}{\theta} \left(\frac{\partial \theta}{\partial \mathbf{F}} \right)_{\mathbf{T}} \right\rangle \right.
$$
$$
\left. + \langle f \rangle \left[\frac{1}{\theta} \left(\frac{\partial \theta}{\partial \mathbf{F}} \right)_{\mathbf{T}} \right] - \left[\left(\frac{\partial f}{\partial \mathbf{F}} \right)_{\mathbf{T}} \right] \right) \cdot \mathbf{N}.
$$

(3.61)

This relation opens the way to determine the velocity of the discontinuity.

3.6   Velocity of moving discontinuity

Having the expression for the jump of the stress tensor in terms of the driving force (3.61), we can relate the driving force to the velocity of the discontinuity.

We return to the jump relation for linear momentum (2.24)

$$
\bar{V}_N [\rho_0 \mathbf{v}] + \mathbf{N} \cdot [\mathbf{T}] = 0.
$$

(3.62)

The application of the Maxwell-Hadamard lemma gives [Maugin and Trimarco (1995)]

$$
[\mathbf{v}] = -[\mathbf{F} \cdot \mathbf{N}] \bar{V}_N,
$$

(3.63)

and the jump relation for linear momentum (3.62) can be rewritten in a form that is more convenient for the calculation of velocity at the discontinuity

$$
\rho_0 \bar{V}_N^2 [\mathbf{F} \cdot \mathbf{N}] = \mathbf{N} \cdot [\mathbf{T}].
$$

(3.64)

Substituting Eq. (3.61) into the jump relation for linear momentum (3.64), we have

$$
\rho_0 \bar{V}_N^2 [\mathbf{F} \cdot \mathbf{N}] = \mathbf{N} \cdot \left(f_S \left\langle \frac{1}{\theta} \left(\frac{\partial \theta}{\partial \mathbf{F}} \right)_{\mathbf{T}} \right\rangle \right.
$$
$$
\left. + \langle f \rangle \left[\frac{1}{\theta} \left(\frac{\partial \theta}{\partial \mathbf{F}} \right)_{\mathbf{T}} \right] - \left[\left(\frac{\partial f}{\partial \mathbf{F}} \right)_{\mathbf{T}} \right] \right).
$$

(3.65)

If we can relate the jumps $[\mathbf{F} \cdot \mathbf{N}]$ and $\mathbf{N} \cdot [\mathbf{T}]$ by means of constitutive equations, we can obtain a kinetic relation taking into account the expression of the jump of normal component of the Piola-Kirchhoff stress tensor in terms of the driving force. This will be illustrated later on.

3.7   Concluding remarks

In summary, the thermomechanical description of moving discontinuities in solids needs to be complemented by additional constitutive information related to the motion of the discontinuity. The usual way to overcome the constitutive deficiency is to introduce a kinetic relation [Abeyaratne and Knowles (1990, 1991, 1994b, 1997a); Abeyaratne, Bhattacharya and Knowles (2001)]. The kinetic relation connecting the driving force to the velocity of the discontinuity is a manifestation of the irreversibility of the process. That is why we apply the local equilibrium jump relations at the moving discontinuity. These relations allow us to determine the jump of the stress excess across the moving discontinuity. But it is still not sufficient to determine the velocity of the discontinuity completely. Therefore, we need to introduce an additional assumption concerning the entropy production at the discontinuity. The simplest assumption of the continuity of the stress excess across the discontinuity is applied. This assumption can be a subject of further modifications or generalizations. Since we use the general material setting, the constitutive behavior of a material is not specified and, therefore, the derived kinetic relation is independent of the constitutive behavior of a material.

The derived local equilibrium jump relations are different for true and quasi-inhomogeneities. They are formulated in terms of excess quantities. The excess quantities are also used instead of numerical fluxes in the proposed numerical algorithm for wave and front propagation, providing a conservative finite-volume numerical scheme. Details of the algorithm description and results of numerical simulations are given in the forthcoming chapters.

Chapter 4

Linear Thermoelasticity

Thermomechanics in the material formulation presented in previous chapters provides a general framework for the description of wave and front propagation in inhomogeneous solids. However, a simpler small-strain approximation is used below for numerical simulations. Therefore, it is instructive to give the necessary expressions in the case of linear thermoelasticity.

4.1 Local balance laws

The free energy per unit volume in *linear isotropic inhomogeneous thermoelasticity* is given by (cf., e.g. [Maugin and Berezovski (1999)])

$$
\begin{aligned}
W\left(\varepsilon_{ij}, \theta; \mathbf{x}\right) = & \frac{1}{2}\left(\lambda(\mathbf{x})\varepsilon_{kk}^2 + 2\mu(\mathbf{x})\varepsilon_{ij}\varepsilon_{ij}\right) \\
& -\frac{C(\mathbf{x})}{2\theta_0}\left(\theta - \theta_0\right)^2 + m(\mathbf{x})\left(\theta - \theta_0\right)\varepsilon_{kk},
\end{aligned}
\tag{4.1}
$$

with the strain tensor in the small-strain approximation

$$
\varepsilon_{ij} = \frac{1}{2}\left(\frac{\partial u_i}{\partial x_j} + \frac{\partial u_j}{\partial x_i}\right),
\tag{4.2}
$$

while u_i are the components of the elastic displacement, $C(\mathbf{x}) = \rho_0 c$, c is the specific heat at constant stress, θ is temperature, θ_0 is a spatially uniform reference temperature and only small deviations from it are envisaged. The *dilatation coefficient* α is related to the thermoelastic coefficient m, and the Lamé coefficients λ and μ by $m = -\alpha(3\lambda + 2\mu)$. The indicated explicit dependence on the point $\mathbf{x}$ means that the body is materially inhomogeneous in general. In particular, we can apply different material parameters for distinct phases or parts of the body.

Neglecting geometrical nonlinearities, the two main equations of thermoelasticity are the *local balance of momentum* at each regular material point in the absence of body force [Nowacki (1986)]:

$$\rho_0(\mathbf{x})\frac{\partial v_i}{\partial t} - \frac{\partial \sigma_{ij}}{\partial x_j} = 0, \tag{4.3}$$

and the *heat propagation equation*

$$\theta\frac{\partial S}{\partial t} + \frac{\partial q_i}{\partial x_i} = 0, \tag{4.4}$$

where t is time, x_j are spatial coordinates, v_i are components of the velocity vector, σ_{ij} is the Cauchy stress tensor, ρ_0 is the density, S is the entropy per unit volume, q_i are components of the heat flux vector.

The entropy and the stress tensor can be expressed in terms of the free energy per unit volume $W = W\left(\varepsilon_{ij}, \theta; \mathbf{x}\right)$ as follows

$$\sigma_{ij} = \frac{\partial W}{\partial \varepsilon_{ij}}, \qquad S = -\frac{\partial W}{\partial \theta}. \tag{4.5}$$

Simultaneously, we assume the Fourier law of heat conduction

$$q_i = -k(\mathbf{x})\frac{\partial \theta}{\partial x_i}, \tag{4.6}$$

where k is the thermal conductivity. Here we could invoke a short digression on isothermal and adiabatic elasticity coefficients (see, for instance, [Maugin (1988)]).

We can now rewrite the relevant *bulk* equations of inhomogeneous linear isotropic thermoelasticity as the following three equations of which the second one is none other than the time derivative of the Duhamel-Neumann thermoelastic constitutive equation [Berezovski, Engelbrecht, and Maugin (2000)]:

$$\rho_0(\mathbf{x})\frac{\partial v_i}{\partial t} = \frac{\partial \sigma_{ij}}{\partial x_j}, \tag{4.7}$$

$$\frac{\partial \sigma_{ij}}{\partial t} = \lambda(\mathbf{x})\frac{\partial v_k}{\partial x_k}\delta_{ij} + \mu(\mathbf{x})\left(\frac{\partial v_i}{\partial x_j} + \frac{\partial v_j}{\partial x_i}\right) + m(\mathbf{x})\frac{\partial \theta}{\partial t}\delta_{ij}, \tag{4.8}$$

$$C(\mathbf{x})\frac{\partial \theta}{\partial t} = \frac{\partial}{\partial x_i}\left(k(\mathbf{x})\frac{\partial \theta}{\partial x_i}\right) + m(\mathbf{x})\frac{\partial v_k}{\partial x_k}. \tag{4.9}$$

The given form of governing equations is convenient for numerical simulation.

4.2 Balance of pseudomomentum

In addition to the governing Eqs. (4.7)-(4.9), it is also instructive to represent the balance of pseudomomentum in the small-strain approximation keeping in mind further considerations concerning the discontinuity propagation. The corresponding equation reads (cf. Eq. (2.48))

$$\frac{\partial \mathcal{P}_i}{\partial t} - \frac{\partial b_{ij}}{\partial x_j} = f_i^{inh} + f_i^{th}, \tag{4.10}$$

where

$$\mathcal{P}_i = -\rho_0 \frac{\partial u_j}{\partial t} \frac{\partial u_j}{\partial x_i} \tag{4.11}$$

is the expression of the pseudomomentum in the small-strain approximation,

$$b_{ij} = -\left(\mathcal{L}\delta_{ij} + \sigma_{jk} \frac{\partial u_k}{\partial x_i} \right) \tag{4.12}$$

is the dynamical Eshelby stress tensor,

$$\mathcal{L} = \frac{1}{2}\rho_0 \frac{\partial u_i}{\partial t} \frac{\partial u_i}{\partial t} - W \tag{4.13}$$

is the Lagrangian density of energy per unit volume, f_i^{inh} is the inhomogeneity force, and f_i^{th} is the thermal force.

For an irreversible process of discontinuity propagation we should take into account the entropy inequality

$$\frac{\partial S}{\partial t} + \frac{\partial (q_i/\theta)}{\partial x_i} \geq 0. \tag{4.14}$$

4.3 Jump relations

Accordingly, we envision the exploitation of the unsteady jump relations established for a moving material discontinuity, but adapt these to the present situation. The jump relation across a discontinuity front S corresponding to the balance of linear momentum (4.3) reads

$$\bar{V}_N[\rho_0 v_i] + N_j[\sigma_{ij}] = 0. \tag{4.15}$$

Both Eqs. (4.10) and (4.14) are nonconservative. Therefore, the corresponding jump relations across the discontinuity front S should exhibit source terms [Maugin (1997)]:

$$\bar{V}_N[\mathcal{P}_i] + N_j[b_{ij}] = -f_i, \tag{4.16}$$

$$\bar{V}_N[S] + N_i \left[\frac{k}{\theta} \frac{\partial \theta}{\partial x_i} \right] = \sigma_S \geq 0. \tag{4.17}$$

Here $\bar{V}_N$ is the normal speed of the points of S, σ_S is the entropy production at the singularity, and f_i are components of an unknown material force. These quantities are constrained to satisfy the second law of thermodynamics at the discontinuity front S such that [Maugin (1997)]

$$f_i V_i = f_S \bar{V}_N = \theta_S \sigma_S \geq 0, \tag{4.18}$$

where θ_S is the temperature at S and f_S is the scalar value of the driving force at the discontinuity

$$f_S = -[W] + \langle \sigma_{ij} \rangle [\varepsilon_{ij}]. \tag{4.19}$$

As previously,

$$[A] = A^+ - A^- \tag{4.20}$$

denotes the jump of the enclosure at S, and $A^\pm$ denote the uniform limits of A in approaching S from the $\pm$ side along the unit normal N_j. The latter is oriented from the "minus" to the "plus" side.

The system of Eqs. (4.7)-(4.9) together with the jump relations (4.15)-(4.17) is used for the numerical simulations of wave and front propagation in the next chapters. Therefore, we will describe briefly the applied numerical algorithm.

4.4   Wave-propagation algorithm: an example of finite volume methods

4.4.1   *One-dimensional elasticity*

For the sake of simplicity, we demonstrate the main features of the algorithm on the simple example of one-dimensional elasticity. In the one-dimensional case, the balance of linear momentum (4.7) is reduced to

$$\rho_0(x) \frac{\partial v}{\partial t} - \frac{\partial \sigma}{\partial x} = 0, \tag{4.21}$$

and the kinematic compatibility condition

$$\frac{\partial \varepsilon}{\partial t} - \frac{\partial v}{\partial x} = 0, \tag{4.22}$$

complemented by Hooke's law

$$\sigma = (\lambda + 2\mu)\varepsilon, \tag{4.23}$$

lead to the reduced equation of motion (4.8) in the form

$$\frac{\partial \sigma}{\partial t} - \rho_0 c_0^2 \frac{\partial v}{\partial x} = 0, \tag{4.24}$$

where $v(x,t)$ is the particle velocity, $\sigma(x,t)$ is the uniaxial stress, $\varepsilon(x,t)$ is a measure of the uniaxial strain, and $c_0 = \sqrt{(\lambda + 2\mu)/\rho_0}$ is the corresponding longitudinal wave velocity.

The system of Eqs. (4.21) - (4.22) can be expressed in the form of a conservation law

$$\frac{\partial}{\partial t}\mathbf{q}(x,t) + \mathbf{A}\frac{\partial}{\partial x}\mathbf{q}(x,t) = 0, \tag{4.25}$$

with

$$\mathbf{q}(x,t) = \begin{pmatrix} \varepsilon \\ \rho_0 v \end{pmatrix} \quad \text{and} \quad \mathbf{A} = \begin{pmatrix} 0 & -1/\rho_0 \\ -\rho_0 c_0^2 & 0 \end{pmatrix}. \tag{4.26}$$

The system of Eqs. (4.21)-(4.22) is hyperbolic, hence the matrix $\mathbf{A}$ is diagonalizable, i.e.

$$\mathbf{R}^{-1}\mathbf{A}\mathbf{R} = \mathbf{\Lambda}, \tag{4.27}$$

where $\mathbf{R}$ is the eigenvector matrix and $\mathbf{\Lambda} = \mathrm{diag}(\lambda^1, \lambda^2)$. In the case of linear elasticity, the eigenvector matrix and its inverse are

$$\mathbf{R} = \begin{pmatrix} 1 & 1 \\ Z & -Z \end{pmatrix}, \quad \mathbf{R}^{-1} = \frac{1}{2Z}\begin{pmatrix} Z & 1 \\ Z & -1 \end{pmatrix}, \tag{4.28}$$

where $Z = \rho_0 c_0$ is an impedance.

It is easy to see that the conservation law (4.25) can be represented as

$$\mathbf{R}^{-1}\frac{\partial}{\partial t}\mathbf{q}(x,t) + \mathbf{R}^{-1}\mathbf{A}\mathbf{R}\mathbf{R}^{-1}\frac{\partial}{\partial x}\mathbf{q}(x,t) = 0, \tag{4.29}$$

and then rewritten in the characteristic form

$$\frac{\partial}{\partial t}\mathbf{w} + \mathbf{\Lambda}\frac{\partial}{\partial x}\mathbf{w} = 0, \tag{4.30}$$

where $\mathbf{w} = \mathbf{R}^{-1}\mathbf{q}$ is introduced. The system of Eqs. (4.30) consists of two decoupled equations for the components of the vector $\mathbf{w}$

$$\frac{\partial w^I}{\partial t} - c_0 \frac{\partial w^I}{\partial x} = 0, \quad \frac{\partial w^{II}}{\partial t} + c_0 \frac{\partial w^{II}}{\partial x} = 0, \tag{4.31}$$

solution of which are left-going and right-going waves

$$w^I(x,t) = w^I(x + c_0 t), \quad w^{II}(x,t) = w^{II}(x - c_0 t). \tag{4.32}$$

Therefore, the general solution of the system of Eqs. (4.25) is the linear combination of these waves by means of eigenvectors

$$\mathbf{q}(x,t) = \mathbf{R}\mathbf{w}(x,t) = w^I(x,t) \begin{pmatrix} 1 \\ Z \end{pmatrix} + w^{II}(x,t) \begin{pmatrix} 1 \\ -Z \end{pmatrix}. \tag{4.33}$$

4.4.2 Averaged quantities

Let us introduce a computational grid of cells $C_n = [x_n, x_{n+1}]$ with interfaces $x_n = n\Delta x$ and time levels $t_k = k\Delta t$. For simplicity, the grid size Δx and time step Δt are assumed to be constant. Integration of Eq. (4.25) over the control volume $C_n \times [t_k, t_{k+1}]$ gives

$$\int_{\Delta x} \mathbf{q}(x, t_{k+1}) dx - \int_{\Delta x} \mathbf{q}(x, t_k) dx$$
$$+ \int_{t_k}^{t_{k+1}} \mathbf{A}\mathbf{q}(x_{n+1}, t) dt - \int_{t_k}^{t_{k+1}} \mathbf{A}\mathbf{q}(x_n, t) dt = 0. \tag{4.34}$$

Equation (4.34) can be rewritten as a numerical scheme in the flux-differencing form

$$\mathbf{Q}_n^{k+1} = \mathbf{Q}_n^k - \frac{\Delta t}{\Delta x}(\mathbf{F}_{n+1}^k - \mathbf{F}_n^k) \tag{4.35}$$

after introducing the average $\mathbf{Q}_n$ of the exact solution on C_n at time $t = t_k$ and the numerical flux $\mathbf{F}_n$ that approximates the time average of the exact flux taken at the interface between the cells C_{n-1} and C_n, i.e.

$$\mathbf{Q}_n \approx \frac{1}{\Delta x} \int_{x_n}^{x_{n+1}} \mathbf{q}(x, t_k) dx, \quad \mathbf{F}_n \approx \frac{1}{\Delta t} \int_{t_k}^{t_{k+1}} \mathbf{A}\mathbf{q}(x_n, t) dt. \tag{4.36}$$

In general, however, the time integrals on the right-hand side of Eq. (4.34) cannot be evaluated exactly since $\mathbf{q}(x_n, t)$ varies with time along each edge of the cell. A fully discrete method follows from an approximation of this average flux based on the values $\mathbf{Q}_n$.

4.4.3 Numerical fluxes

In the Godunov-type finite volume methods, numerical fluxes $\mathbf{F}_n$ are determined by means of the solution of the Riemann problem at interfaces between cells. The solution of the Riemann problem (at the interface between cells $n-1$ and n) consists of two waves, which we denote, following [LeVeque (2002a)], $\mathcal{W}_n^I$ and $\mathcal{W}_n^{II}$. The left-going wave $\mathcal{W}_n^I$ moves into cell $n-1$, the right-going wave $\mathcal{W}_n^{II}$ moves into cell n. The state between the two waves must be continuous across the interface (Rankine-Hugoniot condition) [LeVeque (2002a)]:

$$\mathcal{W}_n^I + \mathcal{W}_n^{II} = \mathbf{Q}_n - \mathbf{Q}_{n-1}. \tag{4.37}$$

In the linear case, the considered waves are determined by eigenvectors of the matrix $\mathbf{A}$ [LeVeque (2002a)]:

$$\mathcal{W}_n^I = \gamma_n^I \mathbf{r}_{n-1}^I, \quad \mathcal{W}_n^{II} = \gamma_n^{II} \mathbf{r}_n^{II}. \tag{4.38}$$

This means that Eq. (4.37) is represented as

$$\gamma_n^I \mathbf{r}_{n-1}^I + \gamma_n^{II} \mathbf{r}_n^{II} = \mathbf{Q}_n - \mathbf{Q}_{n-1}. \tag{4.39}$$

Substituting the eigenvectors into Eq. (4.39), we have

$$\gamma_n^I \begin{pmatrix} 1 \\ Z_{n-1} \end{pmatrix} + \gamma_n^{II} \begin{pmatrix} 1 \\ -Z_n \end{pmatrix} = \mathbf{Q}_n - \mathbf{Q}_{n-1}, \tag{4.40}$$

or, more explicitly,

$$\begin{pmatrix} 1 & 1 \\ \rho_{n-1}c_{n-1} & -\rho_n c_n \end{pmatrix} \begin{pmatrix} \gamma_n^I \\ \gamma_n^{II} \end{pmatrix} = \begin{pmatrix} \bar{\varepsilon}_n - \bar{\varepsilon}_{n-1} \\ \rho\bar{v}_n - \rho\bar{v}_{n-1} \end{pmatrix}. \tag{4.41}$$

Solving the system of linear Eqs. (4.41), we obtain the amplitudes of left-going and right-going waves. Then the numerical fluxes in the Godunov numerical scheme are determined as follows

$$\mathbf{F}_{n+1}^k = -\lambda_{n+1}^I \mathcal{W}_{n+1}^I = -c_{n+1}\gamma_{n+1}^I \mathbf{r}_n^I, \tag{4.42}$$

$$\mathbf{F}_n^k = \lambda_n^{II} \mathcal{W}_n^{II} = -c_n \gamma_n^{II} \mathbf{r}_n^{II}. \tag{4.43}$$

Finally, the Godunov scheme is expressed in the form

$$\mathbf{Q}_n^{k+1} = \mathbf{Q}_n^k + \frac{\Delta t}{\Delta x} \left(c_{n+1}\gamma_{n+1}^I \mathbf{r}_n^I - c_n \gamma_n^{II} \mathbf{r}_n^{II} \right). \tag{4.44}$$

This is the standard form for the wave-propagation algorithm [LeVeque (2002a)].

4.4.4   *Second order corrections*

The scheme considered above is formally first order accurate only. The Godunov scheme exhibits strong numerical dissipation, and discontinuities in the solution are smeared causing low accuracy. In order to increase the order of accuracy, correction terms are introduced [LeVeque (2002a)]. The obtained Lax-Wendroff scheme, on the other hand, is more accurate in smooth parts of the solution, but near discontinuities, numerical dispersion generates oscillations also reducing the accuracy. A successful approach to suppress these oscillations is to apply flux limiters [LeVeque (1997); LeVeque (1998); Fogarthy and LeVeque (1999); Langseth and LeVeque (2000)].

4.4.5   *Conservative wave propagation algorithm*

Another possibility to increase the accuracy on smooth solutions is the conservative wave propagation algorithm [Bale et al. (2003)]. Here the solution of the generalized Riemann problem is obtained by means of the decomposition of the flux difference $\mathbf{A}_n \mathbf{Q}_n - \mathbf{A}_{n-1} \mathbf{Q}_{n-1}$ instead of the decomposition (4.37)

$$\mathcal{L}_n^I + \mathcal{L}_n^{II} = \mathbf{A}_n \mathbf{Q}_n - \mathbf{A}_{n-1} \mathbf{Q}_{n-1}. \tag{4.45}$$

The fluxes $\mathcal{L}^I$ and $\mathcal{L}^{II}$ are still proportional to the eigenvectors of matrix $\mathbf{A}$

$$\mathcal{L}_n^I = \beta_n^I \mathbf{r}_{n-1}^I, \quad \mathcal{L}_n^{II} = \beta_n^{II} \mathbf{r}_n^{II}, \tag{4.46}$$

and the corresponding numerical scheme has the form

$$\mathbf{Q}_n^{l+1} - \mathbf{Q}_n^l = -\frac{\Delta t}{\Delta x} \left(\mathcal{L}_n^{II} + \mathcal{L}_{n+1}^I \right). \tag{4.47}$$

Coefficients β^I and β^{II} are determined from the solution of the system of linear equations

$$\begin{pmatrix} 1 & 1 \\ \rho_{n-1}c_{n-1} & -\rho_n c_n \end{pmatrix} \begin{pmatrix} \beta_n^I \\ \beta_n^{II} \end{pmatrix} = \begin{pmatrix} -(\bar{v}_n - \bar{v}_{n-1}) \\ -(\rho c^2 \bar{\varepsilon}_n - \rho c^2 \bar{\varepsilon}_{n-1}) \end{pmatrix}. \tag{4.48}$$

As it is shown [Bale et al. (2003)], the obtained algorithm is conservative and second-order accurate on smooth solutions.

The wave-propagation method was successfully applied to the simulation of wave propagation in inhomogeneous media with rapidly-varying properties with some additional modifications to ensure the full second order accuracy [LeVeque (1997); LeVeque (1998); Fogarthy and LeVeque

(1999)]. The advantages of the wave-propagation algorithm are high-resolution [Bale et al. (2003)] and the possibility for a natural extension to higher dimensions [Langseth and LeVeque (2000)].

4.5 Local equilibrium approximation

Wave and front propagation in solid mechanics are characterized by the values of velocity of the order of $1000 \, \text{m/s}$. The corresponding characteristic time is of the order of hundreds or even tens of microseconds, especially in impact induced events. It is difficult to expect that the corresponding states of material points during such fast processes are equilibrium ones. The hypothesis of local equilibrium is commonly used to avoid the troubles with non-equilibrium states.

This hypothesis supposes a projection of a non-equilibrium state of a system on a state of an accompanying local equilibrium process [Kestin (1992)]. In finite-volume numerical methods such a projection is achieved by the averaging over the computational cell. This approximation provides the values of local equilibrium parameters in the cell which are in general different from the values of the contact quantities characterizing the non-equilibrium state of the cell by the exchanges. As it was mentioned in chapter 3, the excess quantities should be taken into account. This means that the value of any extensive quantity A is the sum of its averaged counterpart $\bar{A}$ and its excess part A_{ex},

$$A = \bar{A} + A_{ex}. \tag{4.49}$$

4.5.1 *Excess quantities and numerical fluxes*

As we have seen in chapter 3, the decomposition of the free energy density (3.40) in each finite volume element into the averaged (local equilibrium) free energy $\bar{W}$ and the excess free energy W_{ex}

$$W = \bar{W} + W_{ex}, \tag{4.50}$$

immediately reflects in the corresponding decomposition of the stress tensor

$$\sigma = \bar{\sigma} + \Sigma, \tag{4.51}$$

and, by duality, in the decomposition of velocity

$$v = \bar{v} + \mathcal{V}. \tag{4.52}$$

Integration of the governing equations of linear elasticity (4.21)-(4.22) over the computational cell gives

$$\frac{\partial}{\partial t} \int_{\triangle x} \varepsilon dx = v^+ - v^- = \bar{v} + \mathcal{V}^+ - \bar{v} - \mathcal{V}^- = \mathcal{V}^+ - \mathcal{V}^-, \qquad (4.53)$$

$$\rho_0 \frac{\partial}{\partial t} \int_{\triangle x} v dx = \sigma^+ - \sigma^- = \bar{\sigma} + \Sigma^+ - \bar{\sigma} - \Sigma^- = \Sigma^+ - \Sigma^-, \qquad (4.54)$$

where superscripts "+" and "−" denote values of the excess quantities at right and left boundaries of the cell, respectively.

The definition of averaged quantities

$$\bar{\varepsilon} = \frac{1}{\triangle x} \int_{\triangle x} \varepsilon dx, \quad \bar{v} = \frac{1}{\triangle x} \int_{\triangle x} v dx, \qquad (4.55)$$

allows us to rewrite the finite-volume numerical scheme (4.35) in terms of excess quantities

$$(\rho \bar{v})_n^{k+1} - (\rho \bar{v})_n^k = \frac{\Delta t}{\Delta x} \left(\Sigma_n^+ - \Sigma_n^- \right), \qquad (4.56)$$

$$\bar{\varepsilon}_n^{k+1} - \bar{\varepsilon}_n^k = \frac{\Delta t}{\Delta x} \left(\mathcal{V}_n^+ - \mathcal{V}_n^- \right). \qquad (4.57)$$

Though excess quantities are determined formally everywhere inside computational cells, we need to know only their values at the boundaries of the cells, where they play the role of numerical fluxes.

To determine the values of excess quantities at the boundaries between computational cells, we apply the local equilibrium jump relations derived in chapter 3. The excess stress Σ is related to the averaged stress by the local equilibrium jump relation in bulk (3.47), which is reduced in the isothermal case to

$$[\bar{\sigma} + \Sigma] = 0. \qquad (4.58)$$

The same condition follows from the jump relation for the linear momentum (4.15), because the boundary between computational cells does not move. Similarly, the jump relation following from the kinematic compatibility (4.22) reads

$$[\bar{v} + \mathcal{V}] = 0. \qquad (4.59)$$

It is instructive to represent the local equilibrium jump relation (4.58) in the numerical form

$$(\Sigma^+)_{n-1} - (\Sigma^-)_n = (\bar{\sigma})_n - (\bar{\sigma})_{n-1}, \qquad (4.60)$$

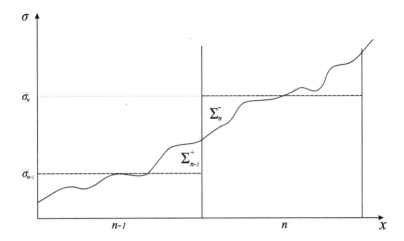

Fig. 4.1 Stresses in the bulk.

which is illustrated in Fig. 4.1. This means that the jump condition (4.58) can be considered as the *continuity of genuine fluxes* at the boundaries between computational cells.

The values of excess stresses and excess velocities at the boundaries between computational cells are not independent. To demonstrate that, we consider the solution of the Riemann problem at the interface between cells.

4.5.2 Riemann problem

The *Riemann problem* at the boundary between computational cells consists of piecewise constant initial data for the system of Eqs. (4.21)-(4.22)

$$\begin{cases} v(x) = \bar{v}_{n-1}, & \sigma(x) = \bar{\sigma}_{n-1}, \text{ if } x < x_n; \\ v(x) = \bar{v}_n, & \sigma(x) = \bar{\sigma}_n, \quad\;\; \text{ if } x > x_n. \end{cases} \tag{4.61}$$

The solution of the Riemann problem is constructed by means of values of Riemann invariants. At the boundary between cells the Riemann invariants hold their values computed by initial data (4.61)

$$\rho c v + \sigma = \rho_n c_n \bar{v}_n + \bar{\sigma}_n, \tag{4.62}$$

$$\rho c v - \sigma = \rho_{n-1} c_{n-1} \bar{v}_{n-1} - \bar{\sigma}_{n-1}. \tag{4.63}$$

Inserting the relations (4.51) and (4.52) into the expressions for the Riemann invariants at the interface (4.62), (4.63) we obtain

$$\rho c v + \sigma = \rho_n c_n \bar{v}_n + \rho_n c_n \mathcal{V}_n^- + \bar{\sigma}_n + \Sigma_n^- = \rho_n c_n \bar{v}_n + \bar{\sigma}_n, \qquad (4.64)$$

$$\rho c v - \sigma = \rho_{n-1} c_{n-1} \bar{v}_{n-1} + \rho_{n-1} c_{n-1} \mathcal{V}_n^+ - \bar{\sigma}_{n-1} - \Sigma_{n-1}^+$$
$$= \rho_{n-1} c_{n-1} \bar{v}_{n-1} - \bar{\sigma}_{n-1}. \qquad (4.65)$$

The latter means that

$$\rho_n c_n \mathcal{V}_n^- + \Sigma_n^- \equiv 0, \qquad (4.66)$$

$$\rho_{n-1} c_{n-1} \mathcal{V}_{n-1}^+ - \Sigma_{n-1}^+ \equiv 0, \qquad (4.67)$$

i.e., the excess quantities depend on each other at the cell boundary.

4.5.3   *Excess quantities at the boundaries between cells*

Rewriting the jump relations (4.58), (4.59) in the numerical form

$$(\Sigma^+)_{n-1} - (\Sigma^-)_n = (\bar{\sigma})_n - (\bar{\sigma})_{n-1}, \qquad (4.68)$$

$$(\mathcal{V}^+)_{n-1} - (\mathcal{V}^-)_n = (\bar{v})_n - (\bar{v})_{n-1}, \qquad (4.69)$$

and using the dependence between excess quantities (Eqs. (4.66) and (4.67)), we obtain then the system of linear equations for the determination of excess velocities

$$\mathcal{V}_{n-1}^+ - \mathcal{V}_n^- = \bar{v}_n - \bar{v}_{n-1}, \qquad (4.70)$$

$$\mathcal{V}_{n-1}^+ \rho_{n-1} c_{n-1} + \mathcal{V}_n^- \rho_n c_n = \rho_n c_n^2 \bar{\varepsilon}_n - \rho_{n-1} c_{n-1}^2 \bar{\varepsilon}_{n-1}. \qquad (4.71)$$

In matrix notation the latter system of equations has the form

$$\begin{pmatrix} 1 & 1 \\ \rho_{n-1} c_{n-1} & -\rho_n c_n \end{pmatrix} \begin{pmatrix} -\mathcal{V}_{n-1}^+ \\ \mathcal{V}_n^- \end{pmatrix} = \begin{pmatrix} -(\bar{v}_n - \bar{v}_{n-1}) \\ -(\rho c^2 \bar{\varepsilon}_n - \rho c^2 \bar{\varepsilon}_{n-1}) \end{pmatrix}. \qquad (4.72)$$

Comparing the obtained equation with Eq. (4.48), we conclude that

$$\beta_n^I = -\mathcal{V}_{n-1}^+, \quad \beta_n^{II} = \mathcal{V}_n^-. \qquad (4.73)$$

This means the values of excess quantities determined by the local equilibrium jump relations at the boundary between computational cells coincide with the numerical fluxes in the conservative wave-propagation algorithm. Consequently, the conservative wave-propagation algorithm is thermodynamically consistent. Moreover, we have no need to prove the convergence and stability of the algorithm in terms of excess quantities because this is proved for the wave-propagation algorithm [LeVeque (2002a)].

The advantage of the wave-propagation algorithm is that every discontinuity in the parameters is taken into account by solving the Riemann problem at each interface between discrete elements.

4.6 Concluding remarks

Linear thermoelasticity equations presented in this chapter include not only the classical balance of linear momentum and the heat conduction equation, but also jump relations at discontinuities associated with these equations as well as those associated with the balance of pseudomomentum and Clausius-Duhem inequality. This leads to the appearance of a driving force providing a possible motion of a discontinuity. The driving force together with the velocity of the discontinuity determines the entropy production due to the irreversible motion of the discontinuity.

The local equilibrium approximation extended to the finite volume computational cells leads to the appearance of excess quantities which are characteristic of non-equilibrium states. Fortunately, we can use the same excess quantities in the proposed numerical algorithm as a replacement of numerical fluxes and even apply the local equilibrium jump relations to determine their values at boundaries between computational cells.

Various examples of wave and front propagation in inhomogeneous solids are presented in the following chapters devoted to numerical simulations based on the proposed algorithm.

Chapter 5

Wave Propagation in Inhomogeneous Solids

Inhomogeneous solids include layered and randomly reinforced composites, multiphase and polycrystalline alloys, functionally graded materials, ceramics and polymers with certain microstructure, etc. Therefore, it is impossible to present a complete theory of linear and nonlinear wave propagation for a full diversity of possible situations, in so far as geometry, contrast of multiphase properties and loading conditions are concerned.

From a practical point of view, we need to perform numerical calculations. Many numerical methods have been proposed to compute wave propagation in heterogeneous solids, among them the stiffness matrix recursive algorithm [Rokhlin and Wang (2002); Wang and Rokhlin (2004)] and spectral layer element method [Chakraborty and Gopalakrishnan (2003, 2004)] should be mentioned in addition to more common finite element, finite difference, and finite volume methods.

However, success in the simulation of wave propagation does not mean that the same algorithm can be applied in the case of moving discontinuity fronts.

In this chapter we will demonstrate how the finite volume wave-propagation algorithm developed by LeVeque (2002a) and reformulated in terms of excess quantities [Berezovski and Maugin (2001)] can be applied to linear and nonlinear wave propagation in materials with rapidly-varying properties. Later the same algorithm will be applied to phase-transition front propagation and to the dynamics of cracks.

Note that both original and modified algorithms are stable, high-order accurate, and thermodynamically consistent.

5.1   Governing equations

The simplest example of heterogeneous media is a periodic medium composed of materials with different properties. With this in view we consider one-dimensional wave propagation in linear elasticity that is governed by the conservation of linear momentum

$$\rho(x)\frac{\partial v}{\partial t} - \frac{\partial \sigma}{\partial x} = 0, \tag{5.1}$$

and the kinematic compatibility condition

$$\frac{\partial \varepsilon}{\partial t} = \frac{\partial v}{\partial x}. \tag{5.2}$$

The closure of the system of Eqs. (5.1), (5.2) is achieved by a constitutive relation, which in the simplest case is Hooke's law

$$\sigma = \rho(x)c^2(x)\,\varepsilon, \tag{5.3}$$

where $c(x) = \sqrt{(\lambda(x) + 2\mu(x))/\rho(x)}$ is the corresponding longitudinal wave velocity. The indicated explicit dependence on the point x means that the medium is materially inhomogeneous.

The system of Eqs. (5.1) - (5.3) is solved numerically by means of the wave-propagation algorithm described in detail in the foregoing chapter.

5.2   One-dimensional waves in periodic media

As a first example, we consider the propagation of a pulse in a finely periodic medium. The initial form of the pulse is given in Fig. 5.1 where the fine periodic variation in density is also shown by dashed lines.

For the test problem, materials are chosen as polycarbonate ($\rho = 1190$ kg/m^3, $c = 4000$ m/s) and Al 6061 ($\rho = 2703$ kg/m^3, $c = 6149$ m/s).

We apply the numerical scheme (4.56), (4.57) for the solution of the system of equations (5.1)-(5.3). The corresponding excess quantities are calculated by means of Eqs. (4.68)-(4.71).

As already noted, we can exploit all the advantages of the wave-propagation algorithm [LeVeque (1997)]. However, no limiters are used in the calculations. Spurious oscillations are suppressed by means of using a first-order Godunov step after each three second-order Lax-Wendroff steps. This idea of composition was invented by Liska and Wendroff (1998).

Calculations are performed with Courant-Friedrichs-Levy number equal to 1. The result of the simulation for 4000 time steps is shown in Fig. 5.2.

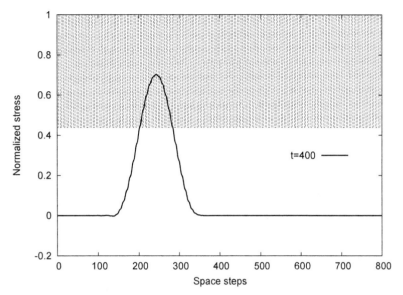

Fig. 5.1   Initial pulse shape.

We observe a distortion of the pulse shape and a decrease in the velocity of the pulse propagation in comparison to the maximal longitudinal wave velocity in the materials. These results correspond to the prediction of the effective media theory by Santosa and Symes (1991) both qualitatively and quantitatively [Fogarthy and LeVeque (1999)].

It should be noted that the effective media theory [Santosa and Symes (1991)] leads to the dispersive wave equation

$$\frac{\partial^2 u}{\partial t^2} = (c^2 - c_a^2)\frac{\partial^2 u}{\partial x^2} + p^2 c_a^2 c_b^2 \frac{\partial^4 u}{\partial x^4}, \tag{5.4}$$

where u is the displacement, p is the periodicity parameter, c_a and c_b are parameters of the effective media [Engelbrecht et al. (2005)], instead of the wave equation following from Eqs. (5.1) - (5.3)

$$\frac{\partial^2 u}{\partial t^2} = c^2 \frac{\partial^2 u}{\partial x^2}. \tag{5.5}$$

Equation (5.4) exhibits *dispersion* (fourth-order space derivative) and the alteration in the longitudinal wave speed.

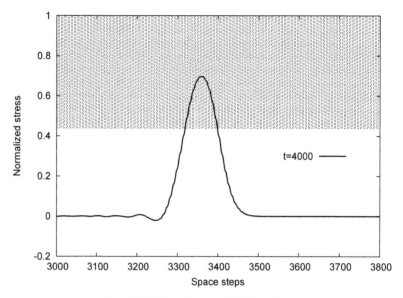

Fig. 5.2   Pulse shape at 4700 time steps.

5.3   One-dimensional weakly nonlinear waves in periodic media

In the next example, we examine the influence of materials nonlinearity on the wave propagation. To close the system of Eqs. (5.1), (5.2) in the case of weakly nonlinear media we apply a simple nonlinear stress-strain relation

$$\sigma = \rho c^2 \varepsilon (1 + A\varepsilon), \qquad (5.6)$$

where A is a parameter of nonlinearity, values and sign of which are supposed to be different for hard and soft materials.

The solution method is almost the same as previously mentioned. The approximate Riemann solver for the nonlinear elastic media (Eq. (5.6)) is similar to that used in LeVeque (2002b); LeVeque and Yong (2003). This means that a modified longitudinal wave velocity, $\hat{c}$, following the nonlinear stress-strain relation (5.6), is applied at each time step in the numerical scheme (4.56), (4.57)

$$\hat{c} = c\sqrt{1 + 2A\varepsilon} \qquad (5.7)$$

instead of the piece-wise constant one corresponding to the linear case.

We consider the same pulse shape and the same materials (polycarbonate and Al 6061) as in the case of the linear periodic medium. However, the nonlinear effects appear only for a sufficiently high magnitude of loading. The values of the parameter of nonlinearity A were chosen as 0.24 for Al 6061 and 0.8 for polycarbonate. The results of simulations corresponding to 400, 1600, and 5200 time steps are shown in Figs. 5.3-5.5.

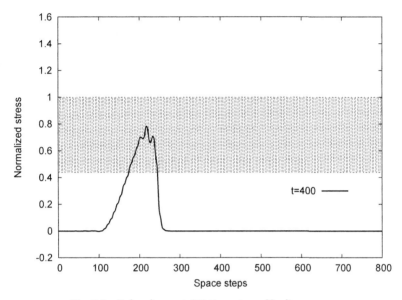

Fig. 5.3   Pulse shape at 400 time steps. Nonlinear case.

We observe that an initial bell-shaped pulse is transformed in a train of soliton-like pulses propagating with amplitude-dependent speeds. This kind of behavior was first reported by LeVeque (2002b), who called these pulses "stegotons" because their shape is influenced by periodicity.

In principle, soliton-like solution could be expected because if we combine the weak nonlinearity (5.6) with the dispersive wave equation in terms of the effective media theory (5.4), we arrive at the Boussinesq-type equation

$$\frac{\partial^2 u}{\partial t^2} = (c^2 - c_a^2)\frac{\partial^2 u}{\partial x^2} + \alpha A \frac{\partial u}{\partial x}\frac{\partial^2 u}{\partial x^2} + p^2 c_a^2 c_b^2 \frac{\partial^4 u}{\partial x^4}, \qquad (5.8)$$

which possesses soliton-like solutions (Maugin, 1999a).

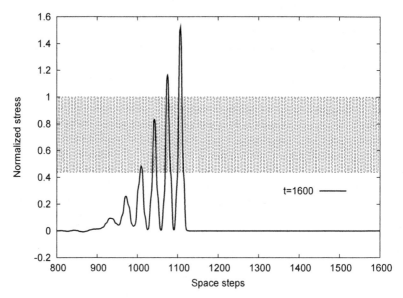

Fig. 5.4　Pulse shape at 1600 time steps. Nonlinear case.

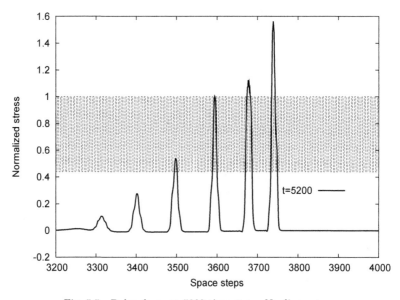

Fig. 5.5　Pulse shape at 5200 time steps. Nonlinear case.

5.4   One-dimensional linear waves in laminates

There are three basic length scales in wave propagation phenomena:

- The typical wavelength λ;
- The typical size of the inhomogeneities d;
- The typical size of the whole inhomogeneity domain L.

In the case of the above-considered infinite periodic media the third length scale was absent. Therefore, it may be instructive to consider wave propagation in a body where the periodic arrangement of layers of different materials is confined within a finite spatial domain.

To investigate the influence of the size of the inhomogeneity domain, we compare the shape of the pulse in the homogeneous medium with the corresponding pulse transmitted through the periodic array with a different number of distinct layers (Fig. 5.6).

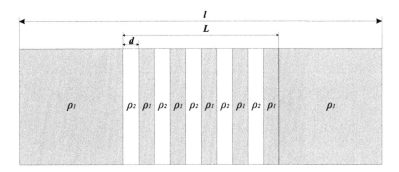

Fig. 5.6   Length scales in laminate.

We use Ti ($\rho = 4510$ kg/m^3, $c = 5020$ m/s) and Al ($\rho = 2703$ kg/m^3, $c = 5240$ m/s) as components in the numerical simulations of linear elastic wave propagation. Results of the comparison are presented in Figs. 5.7-5.9.

As we can see, in the case $d \ll \lambda$ (Fig. 5.7) the behavior of the pulse is almost the same as in the case of infinite periodicity. The increasing of the size of the inhomogeneity domain leads to the pulse delay increasing, in comparison to the pure homogeneous case.

We observe a much more complicated behavior if the size of the periodicity d approaches the wavelength λ (Fig. 5.8). As one can see, the amplitude of the main pulse decreased with the increase in size of the inhomogeneity

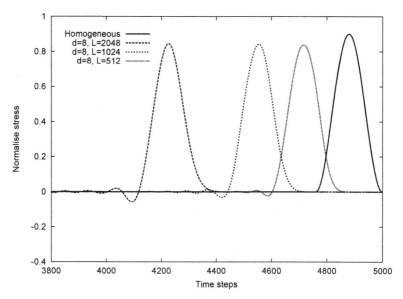

Fig. 5.7   Pulse shape at 5000 time step.

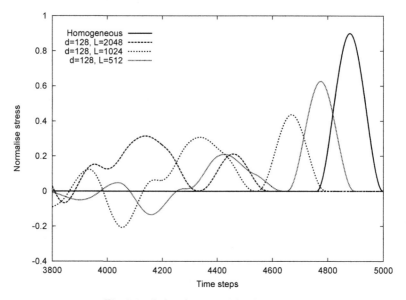

Fig. 5.8   Pulse shape at 5000 time steps.

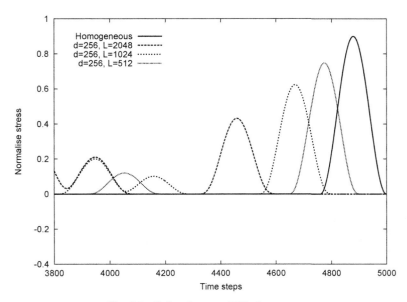

Fig. 5.9   Pulse shape at 5000 time steps.

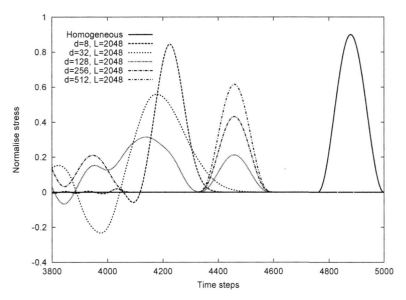

Fig. 5.10   Pulse shape at 5000 time steps.

domain, and secondary waves appear. The situation becomes even clearer in the case of comparable periodicity and initial wavelength (Fig. 5.9).

To analyze the effect of the periodicity size, we compare the shapes of the pulse corresponding to different periodicitiess for the fixed size of the inhomogeneity domain in Fig. 5.10.

In this figure we can observe how the pulse corresponding to small size of periodicity expands into two separate pulses with increasing of the periodicity size. This is nothing but pure dispersion, i.e. the separation of the wave into components of various frequencies.

Thus, waves in laminates demonstrate dispersive behavior which is governed by relations between characteristic length scales. Taking into account nonlinear effects we have witnessed the soliton-like wave propagation. Both nonlinearity and dispersion effects are observed experimentally in laminates under shock loading.

5.5   Nonlinear elastic wave in laminates under impact loading

Though the stress response to an impulsive shock loading is very well understood for homogeneous materials, the same cannot be said for heterogeneous systems. In heterogeneous media, scattering due to interfaces between dissimilar materials plays an important role for shock wave dissipation and dispersion [Grady (1998)].

Diagnostic experiments for the dynamic behavior of heterogeneous materials under impact loading are usually carried out using a plate impact test configuration under a one-dimensional strain state. These experiments were recently reviewed in [Chen and Chandra (2004); Chen, Chandra and Rajendran (2004)]. For almost all these experiments, the stress response has shown a sloped rising part followed by an oscillatory behavior with respect to a mean value [Chen and Chandra (2004); Chen, Chandra and Rajendran (2004)]. Such a behavior in periodically layered systems is consistently exhibited in the systematic experimental work by Zhuang, Ravichandran and Grady (2003). The specimens used in the shock compression experiments [Zhuang, Ravichandran and Grady (2003)] were periodically layered two-component composites prepared by repeating a composite unit as many times as necessary to form a specimen with the desired thickness (see Fig. 5.11).

A buffer layer of the same material as the soft component of the speci-

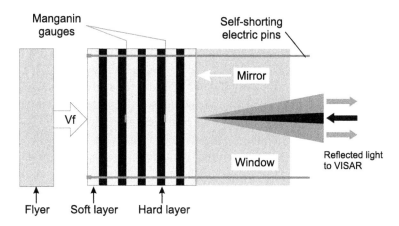

Fig. 5.11 Experimental setting.

men was used at the other side of the specimen. A window in contact with the buffer layer was used to prevent the free surface from serious damage due to unloading from shock wave reflection at the free surface. Shock compression experiments were conducted by employing a powder gun loading system, which could accelerate a flat plate flyer to a velocity in the range of 400 m/s to about 2000 m/s. In order to measure the particle velocity history at the specimen window surface, a velocity interferometry system was constructed, and the manganin stress technique was adopted to measure the shock stress history at selected internal interfaces. Four different materials, polycarbonate, 6061-T6 aluminum alloy, 304 stainless steel, and glass, were chosen as components. The selection of these materials provided a wide range of combinations of shock wave speeds, acoustic impedance and strength levels. The influence of multiple reflections of internal interfaces on shock wave propagation in the layered composites was clearly illustrated by the shock stress profiles measured by manganin gages. The origin of the observed structure of the stress waves was attributed to material heterogeneity at the interfaces. For high velocity impact loading conditions, it was fully realized that material nonlinear effects may play a key role in altering the basic structure of the shock wave.

Amongst the modeling efforts, the mechanical behavior of composites has been extensively investigated using the homogenization approach [Hashin (1983)]. Since this approach does not directly consider the interfaces, it is limited in examining the impact behavior, where wave interactions can be very important.

An approximate solution for layered heterogeneous materials subjected to high velocity plate impact has been developed [Chen and Chandra (2004); Chen, Chandra and Rajendran (2004)]. For laminated systems under shock loading, shock velocity, density and volume were related to the particle velocity via an equation of state. The elastic analysis was extended to shock response by incorporating the nonlinear effects through computing shock velocities of the wave trains and superimposing them.

As pointed out in [Zhuang, Ravichandran and Grady (2003)], stress wave propagation through layered media made of isotropic materials provides an ideal model to investigate the effect of heterogeneous materials under shock loading, because the length scales, e.g., thickness of individual layers, and other measures of heterogeneity, e.g., impedance mismatch, are well defined.

Since the impact velocity in shock experiments is sufficiently high, various nonlinear effects may affect the observed behavior. This is why we apply numerical simulations of finite-amplitude nonlinear wave propagation to the study of scattering, dispersion and attenuation of shock waves in layered heterogeneous materials.

5.5.1   *Problem formulation*

The geometry of the problem follows the experimental configuration [Zhuang, Ravichandran and Grady (2003)] (Fig. 5.12).

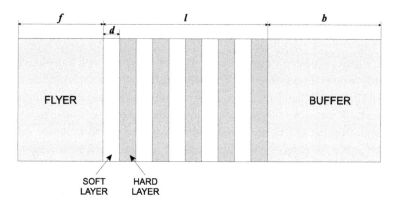

Fig. 5.12   Geometry of the problem.

We consider the initial-boundary value problem of impact loading of a heterogeneous medium composed of alternating layers of two different materials. The impact is provided by a planar flyer of length f which has an initial velocity v_0. A buffer of the same material as the soft component of the specimen is used to eliminate the effect of wave reflection at the stress-free surface. The densities of the two materials are different, and the materials response to compression is characterized by the distinct stress-strain relations $\sigma(\varepsilon)$. Compressional waves propagating in the direction of layering are modeled by the one-dimensional hyperbolic system of conservation laws

$$\rho \frac{\partial v}{\partial t} = \frac{\partial \sigma}{\partial x}, \qquad \frac{\partial \varepsilon}{\partial t} = \frac{\partial v}{\partial x}, \tag{5.9}$$

where $\varepsilon(x, t)$ is the strain and $v(x, t)$ the particle velocity.

Initially, stress and strain are zero inside the flyer, the specimen, and the buffer, but the initial velocity of the flyer is nonzero:

$$v(x, 0) = v_0, \quad 0 < x < f, \tag{5.10}$$

where f is the size of the flyer. The left and right boundaries are both stress-free.

Instead of the equation of state like the one used in [Chen and Chandra (2004); Chen, Chandra and Rajendran (2004)], we apply a simpler nonlinear stress-strain relation $\sigma(\varepsilon, x)$ for each material (Eq. (5.6)) (cf. [Meurer, Qu and Jacobs (2002)])

$$\sigma = \rho c^2 \varepsilon (1 + A\varepsilon), \tag{5.11}$$

where ρ is the density, c is the conventional longitudinal wave speed, A is a parameter of nonlinearity, values and signs of which are supposed to be different for hard and soft materials.

We apply the same numerical scheme as previously discussed. Results of the numerical simulations compared with experimental data [Zhuang, Ravichandran and Grady (2003)] are presented in the next section.

5.5.2 *Comparison with experimental data*

Figure 5.13 shows the measured and calculated stress time history in the composite, which consists of 8 units of polycarbonate, each 0.74 mm thick, and 8 units of stainless steel, each 0.37 mm thick. The material properties of components are extracted from Zhuang, Ravichandran and Grady (2003): the density $\rho = 1190$ kg/cm^3, the sound velocity $c = 1957$ m/s for the polycarbonate and $\rho = 7890$ kg/cm^3, $c = 5744$ m/s for the stainless steel. The stress time histories correspond to the distance 0.76 mm from the

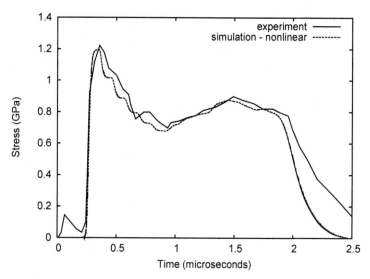

Fig. 5.13   Comparison of shock stress time histories corresponding to the experiment
112501 [Zhuang, Ravichandran and Grady (2003)].

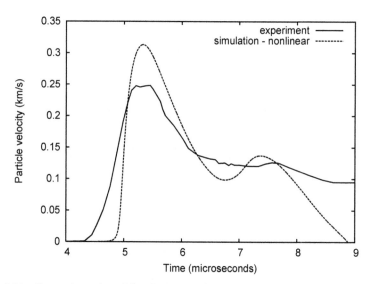

Fig. 5.14   Comparison of particle velocity time histories corresponding to the experiment
112501 [Zhuang, Ravichandran and Grady (2003)].

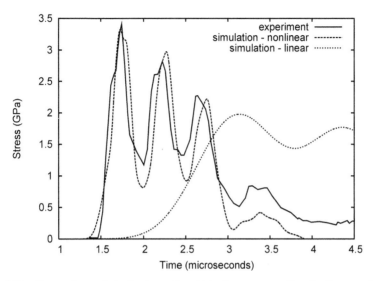

Fig. 5.15 Comparison of shock stress time histories corresponding to the experiment 110501 [Zhuang, Ravichandran and Grady (2003)].

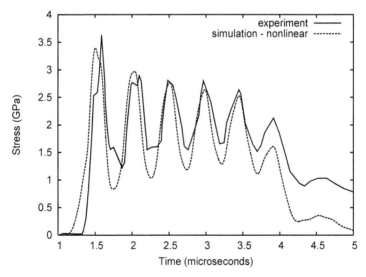

Fig. 5.16 Comparison of shock stress time histories corresponding to the experiment 110502 [Zhuang, Ravichandran and Grady (2003)].

impact face. Calculations are performed for the flyer velocity 561 m/s and the flyer thickness 2.87 mm.

Results of numerical calculations depend crucially on the choice of the parameter of nonlinearity A. We choose this parameter from the conditions to match the numerical simulations to experimental results (see discussion below).

Time histories of particle velocity for the same experiment are shown in Fig. 5.14. It should be noted that the particle velocity time histories correspond to the boundary between the specimen and the buffer. As one can see both stress and particle velocity time histories are well reproduced by the nonlinear model with the same values of the nonlinear parameter A.

As pointed out in Zhuang, Ravichandran and Grady (2003), the influence of multiple reflections of internal interfaces on shock wave propagation in the layered composites is clearly illustrated by the shock stress time histories measured by manganin gages. Therefore, we focus our attention on the comparison of the stress time histories.

Figure 5.15 shows the stress time histories in the composite, which consists of 16 units of polycarbonate, each 0.37 mm thick, and 16 units of stainless steel, each 0.19 mm thick. The stress time histories correspond to the distance 3.44 mm from the impact face. Calculations are performed for the flyer velocity 1043 m/s and the flyer thickness 2.87 mm. The nonlinear parameter A is chosen here to be 2.80 for polycarbonate and zero for stainless steel. Additionally, the stress time history corresponding to the linear elastic solution (i.e., nonlinear parameter A is zero for both components) is shown. It can be seen that the stress time history computed by means of the considered nonlinear model is very close to the experimental one. It reproduces three main peaks and decreases with distortion, as it is observed in the experiment [Zhuang, Ravichandran and Grady (2003)].

In Fig. 5.16 the same comparison is presented for the same composite as in Fig. 5.15, only the flyer thickness is different (5.63 mm). This means that the shock energy is approximately twice higher than that in the previous case. The nonlinear parameter A is also increased to 4.03 for polycarbonate and remains zero for stainless steel. As a result, all 6 experimentally observed peaks are reproduced well.

In Fig. 5.17 the comparison of stress time histories is presented for the composite, which consists of 16 0.37 mm thick units of polycarbonate and 16 0.20 mm thick units of D-263 glass. The material properties of D-263 glass are [Zhuang, Ravichandran and Grady (2003)]: the density $\rho = 2510$ kg/cm^3, the sound velocity $c = 5703$ m/s. The distance between the

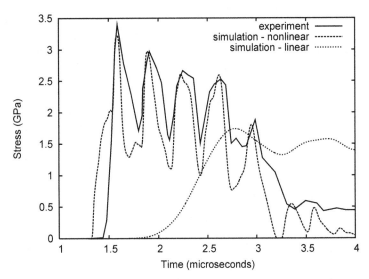

Fig. 5.17   Comparison of shock stress time histories corresponding to the experiment 112301 [Zhuang, Ravichandran and Grady (2003)].

measurement point and the impact face is 3.41 mm. Corresponding flyer velocity is 1079 m/s and the flyer thickness is 2.87 mm. The nonlinear parameter A is chosen to be equal 5.025 for polycarbonate and zero for D-263 glass. Again, the stress time history corresponding to the linear elastic solution (i.e., nonlinear parameter is zero for both components) is shown. As one can see, the stress time history corresponding to the nonlinear model reproduces all 5 peaks with the same amplitude as observed experimentally.

Figure 5.18 shows the comparison of stress time histories for composite, which consists of 7 units of polycarbonate, each 0.74 mm thick, and 7 units of float glass, each 0.55 mm thick. The material properties of float glass are slightly different from those for D-263 glass [Zhuang, Ravichandran and Grady (2003)]: the density $\rho = 2500$ kg/cm^3, the sound velocity $c = 5742$ m/s. The stress profiles correspond to the distance 3.37 mm from the impact face, to the flyer velocity 563 m/s, and to the flyer thickness 2.87 mm. The nonlinear parameter A is equal to 3.04 for polycarbonate and zero for float glass. The result of the numerical simulation coincides with those of the experiment both in amplitude and shape.

In Fig. 5.19 the same comparison is presented for the same composite, only the flyer velocity is almost doubly higher, at 1056 m/s. The value of

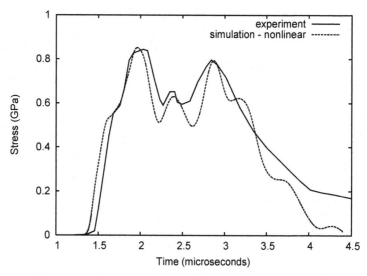

Fig. 5.18   Comparison of shock stress time histories corresponding to the experiment 120201 [Zhuang, Ravichandran and Grady (2003)].

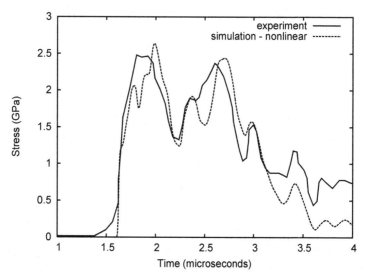

Fig. 5.19   Comparison of shock stress time histories corresponding to the experiment 120202 [Zhuang, Ravichandran and Grady (2003)].

the nonlinear parameter A is 5.53 for polycarbonate and zero for float glass. It can be observed that the result of the numerical simulation is very close to experimental data. The complicated shape of the experimental stress time history is reproduced as well.

As can be seen, the agreement between results of calculations and experiments is achieved by the adjustment of the nonlinear parameter A.

5.5.3 *Discussion of results*

Though the parameter of nonlinearity A looks like a material constant in Eq. (5.11), numerical simulations show this parameter also depends on the structure of the specimen. The nonlinear parameter values, together with the used experimental conditions are given in Table 5.1. In the table, PC denotes polycarbonate, GS - glass, SS - 304 stainless steel; the number following the abbreviation of the component material represents the layer thickness in hundredths of a millimeter.

Table 5.1 Experimental conditions and values of the parameter of nonlinearity.

Exp.	Specimen soft/hard	Units	Flyer velocity (m/s)	Flyer thickness (mm)	Gage position (mm)	A (PC)	A other
110501	PC37/SS19	16	1043	2.87 (PC)	3.44	2.80	0
110502	PC37/SS19	16	1045	5.63 (PC)	3.44	4.03	0
112301	PC37/GS20	16	1079	2.87 (PC)	3.41	5.025	0
120201	PC74/GS55	7	563	2.87 (PC)	3.37	3.04	0
120202	PC74/GS55	7	1056	2.87 (PC)	3.35	5.53	0

It appears that the application of the nonlinear model only to the soft material (polycarbonate) is sufficient to reproduce stress profiles at the gage position about 3.4 mm; any hard material can be treated as a linear elastic one.

The comparison of experimental conditions in experiments 110501 and 110502 as well as 120201 and 120202 and the corresponding values of the parameter of nonlinearity A demonstrates the dependence of the parameter A on the impact energy. The influence of the impedance mismatch clearly follows from the results of simulations corresponding to experiments 110501 and 112301. The dependence on the number of layers is not clear: the difference between the values of the nonlinear parameter in the simulations of experiments 112301 and 120202 can be attributed to the slightly different material properties of float glass and D-263 glass. The effect of the thickness ratio of the layers mentioned in [Chen and Chandra (2004)] cannot be investigated on the basis of the discussed experimental data, since the thickness ratio was unchanged in the experiments by Zhuang, Ravichandran and Grady (2003).

It follows that the nonlinear behavior of the soft material is affected not only by the energy of the impact but also by the scattering induced by internal interfaces. It should be noted that the influence of the nonlinearity is not necessarily small. In the numerical simulations matching the experiments, the increase of the actual sound velocity of polycarbonate follows. It may be up to two times in comparison with the linear case. This conclusion is really surprising but supported by the stress time histories. Such an effect but at smaller scale has also been shown by LeVeque (2002b); LeVeque and Yong (2003).

To compare the results of experiments with different geometry and loading conditions, we need to have a similarity in the experimental setting. However, experimental data in Zhuang, Ravichandran and Grady (2003) correspond to various impact energies, impedance mismatches, and number and thickness of units. Therefore, we need to normalize the experimental conditions. First of all, we choose one of the experiments as a representative one. For example, we can choose experiment marked as 110501 as a representative one. We then relate all other experimental conditions to the conditions of the representative experiment. This means the impact energy for each experiment should be normalized with respect to the impact energy corresponding to the experiment 110501 resulting in the normalized impact energy $\check{E}$. Similarly, the impedance ratio of hard and soft materials should be normalized with respect to the corresponding ratio for the experiment 110501 to obtain the normalized impedance ratio $\check{Z}$. The geometrical factor can be introduced as follows:

$$G = \frac{mh_2}{h_1 + h_2}, \tag{5.12}$$

where m is gage position, h_1 and h_2 are thicknesses of soft and hard layers, respectively. Its normalized value $\check{G}$ is obtained as described above.

We can then compute a modified parameter of nonlinearity $\check{A}$

$$\check{A} = A\sqrt{\frac{\check{Z}}{\check{E}\check{G}}}. \tag{5.13}$$

The results of calculations are given in Table 5.2. As one can see, the modified values of the parameter of nonlinearity deviate from the mean value (equal to 2.806) by less than 3.5 %.

Table 5.2 Normalized experimental conditions and nonlinearity parameters.

Exp.	Specimen soft/hard	Normalized flyer energy $\check{E}$	Relative impedance mismatch $\check{Z}$	Geometrical factor $\check{G}$	A (PC)	$A\sqrt{\frac{\check{Z}}{EG}}$
110501	PC37/SS19	1.00	1.00	1.00	2.80	2.80
110502	PC37/SS19	1.97	1.00	1.00	4.03	2.87
112301	PC37/GS20	1.07	0.316	1.02	5.025	2.71
120201	PC74/GS55	0.29	0.316	1.226	3.04	2.87
120202	PC74/GS55	1.025	0.316	1.226	5.53	2.78

The possibility to calculate the single value of the parameter of nonlinearity means that there exists a similarity in the process under different impact energies, impedance mismatches and geometry. Therefore, the value of the parameter of nonlinearity can be calculated following the simple similarity relation (5.13) from one set of experimental conditions with respect to another.

It should also be noted that the equation of state suggested for the simulation of the plate impact test in [Chen and Chandra (2004); Chen, Chandra and Rajendran (2004)] is simply an approximation of the relation (5.7) in the case of very small deformations. In fact,

$$\hat{c} = c\sqrt{1 + 2A\varepsilon} \sim c(1 + A\varepsilon) \quad \text{for} \quad A\varepsilon \ll 1. \tag{5.14}$$

The nonlinear part $A\varepsilon$ can be represented as Av/c at least under condition $dx/dt = c$, which leads to the equation of state

$$\hat{c} = c + Av, \tag{5.15}$$

mentioned above. This kind of equation of state is also condition-dependent since the particle velocity v depends definitely on the structure of a specimen [Berezovski et al. (2006)].

Thus, the application of a nonlinear stress-strain relation for materials in numerical simulations of the plate impact problem of a layered heterogeneous medium shows that a good agreement between computations and experiments can be obtained by adjusting the values of the parameter of nonlinearity. In the numerical simulations of the finite-amplitude shock wave propagation in heterogeneous composites, the flyer size and velocity, impedance mismatch of hard and soft materials, as well as the number and size of layers in a specimen were the same as in experiments by Zhuang, Ravichandran and Grady (2003). Moreover, a nonlinear behavior of materials was also taken into consideration. This means that combining scattering effects induced by internal interfaces and physical nonlinearity in materials behavior into one nonlinear parameter provides the possibility to reproduce the shock response in heterogeneous media observed experimentally. In this context, parameter A is actually influenced by (i) the physical nonlinearity of the soft material and (ii) the mismatch of elasticity properties of soft and hard materials. The mismatch effect is similar to the type of nonlinearity characteristic to materials with different moduli of elasticity for tension and compression. The mismatch effect manifests itself due to wave scattering at the internal interfaces, and therefore, depends on the structure of a specimen. The variation of the parameter of nonlinearity confirms the statement that the nonlinear wave propagation is highly affected by interaction of the wave with the heterogeneous substructure of a solid [Zhuang, Ravichandran and Grady (2003)].

The relation between different values of the parameter of nonlinearity is found by means of the normalization of experimental conditions. The obtained similarity means that the same physical mechanism can manifest itself differently depending on the particular heterogeneous substructure.

It should be noted that layered media do not exhaust all possible substructures of heterogeneous materials. Another example of heterogeneous substructure is provided by functionally graded materials.

5.6   Waves in functionally graded materials

Functionally graded materials (FGMs) are composed of two or more phases, that are fabricated so that their compositions vary more or less continuously in some spatial direction and are characterized by nonlinear gradients that result in graded properties. Traditional composites are homogeneous mixtures, and they therefore involve a compromise between the desirable properties of the component materials. Since significant proportions of an FGM contain the pure form of each component, the need for compromise is eliminated. The properties of both components can be fully utilized. For example, the toughness of a metal can be mated with the refractoriness of a ceramic, without any compromise in the toughness of the metal side or the refractoriness of the ceramic side.

Comprehensive reviews of current FGM research may be found in the papers by Hirai (1996) and Markworth, Ramesh and Parks (1995), and in the book by Suresh and Mortensen (1998).

Studies of the evolution of stresses and displacements in FGMs subjected to quasistatic loading [Suresh and Mortensen (1998)] show that the utilization of structures and geometry of a graded interface between two dissimilar layers can reduce stresses significantly. Such an effect is also important in case of dynamical loading where energy-absorbing applications are of special interest.

We consider the one-dimensional problem in elastodynamics for an FGM slab in which material properties vary only in the thickness direction. It is assumed that the slab is isotropic and inhomogeneous with the following fairly general properties [Chiu and Erdogan (1999)]:

$$E'(x) = E_0' \left(a\frac{x}{l} + 1 \right)^m, \quad \rho(x) = \rho_0 \left(a\frac{x}{l} + 1 \right)^n, \quad (5.16)$$

where ρ is the mass density, l is the thickness, a, m, and n are arbitrary real constants with $a > -1$, E_0 and ρ_0 are the elastic constant and density at $x = 0$. The elastic constant E_0 is determined under the assumption that $\sigma_{yy} = \sigma_{zz}$ and the slab is fully constrained at infinity. It can thus be shown that

$$E' = \frac{E(1-\nu)}{(1+\nu)(1-2\nu)}, \quad (5.17)$$

where $E(x)$ and $\nu(x)$ being the Young's modulus and the Poisson's ratio of the inhomogeneous material.

It is assumed that the slab is at rest for $t \leq 0$, therefore, the following initial conditions are valid:

$$v(x,0) = 0, \quad \sigma(x,0) = 0. \quad (5.18)$$

The boundary condition at $x = 0$ is

$$v(0,t) = 0, \quad t > 0 \quad (\text{"fixed" boundary}). \quad (5.19)$$

At $x = l$ the slab is subjected to a stress pulse given by

$$\sigma_{xx}(l,t) = \sigma_0 f(t), \quad t > 0, \quad (5.20)$$

where the constant σ_0 is the magnitude of the pulse, the function f describes its time profile, and without any loss in generality, it is assumed that if $|f| \leq 1$.

Following Chiu and Erdogan (1999), we consider an FGM slab that consists of nickel and zirconia. The thickness of the slab is $l = 5$ mm, on one surface the medium is pure nickel, on the other surface pure zirconia, and the material properties $E_0(x)$ and $\rho(x)$ vary smoothly in thickness direction. A pressure pulse defined by

$$\sigma_{xx}(l,t) = \sigma_0 f(t) = -\sigma_0 (H(t) - H(t - t_0)) \quad (5.21)$$

Table 5.3   Properties of materials.

	E (GPa)	ν	ρ (kg/m^3)
ZrO	151	0.33	5331
Ni	207	0.31	8900

is applied to the surface $x = l$. Here H is the Heaviside function. The pulse duration is assumed to be $t_0 = 0.2\,\mu s$. The properties of the constituent materials used are given in Table 5.3 [Chiu and Erdogan (1999)].

The material parameters in Eq. (5.16) for the FGMs used are [Chiu and Erdogan (1999)]: $a = -0.12354$, $m = -1.8866$, and $n = -3.8866$. The stress is calculated up to $12\,\mu s$ (the propagation time of the plane wave through the thickness $l = 5$ mm is approximately $0.77\,\mu s$ in pure ZrO$_2$ and $0.88\,\mu s$ in Ni).

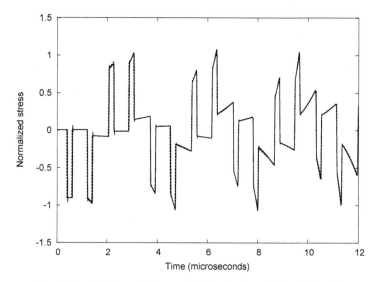

Fig. 5.20   Variation of stress with time in the middle of the slab.

Numerical simulations were performed by means of the same algorithm as above. Comparison of the results of the numerical simulation and the analytical solution by Chiu and Erdogan (1999) for the time dependence of the normalized stress σ_{xx}/σ_0 at the location $x/l = 1/2$ is shown in Fig. 5.20.

As one can see, it is difficult to make a distinction between analytical and numerical results. This means that the applied algorithm is well suited for the simulation of wave propagation in FGM.

A nonlinear behavior for the same materials with the nonlinearity parameter $A=0.19$ is shown in Fig. 5.21. For the comparison, calculations were performed with the value 0.9 of the Courant number in both linear and nonlinear cases.

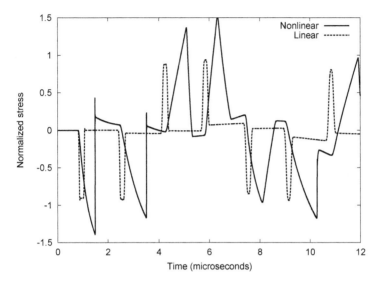

Fig. 5.21 Variation of stress with time in the middle of the slab. Nonlinear case.

The amplitude amplification and pulse shape distortion in comparison with linear case is clearly observed. In addition, the velocity of a pulse in the nonlinear material is increased.

5.7 Concluding remarks

As we have seen, linear and non-linear wave propagation in media with rapidly-varying properties as well as in functionally graded materials can be successfully simulated by means of the modification of the wave propagation algorithm based on the non-equilibrium jump relation for true inhomogeneities. It should be emphasized that the used jump relation express the continuity of genuine unknown fields at the boundaries between computational cells. The applied algorithm is conservative, stable up to Courant number equal to 1, high-order accurate, and thermodynamically consistent. However, the main advantage of the presented modification of the wave-propagation algorithm is its applicability to the simulation of moving discontinuities. This property is related to the formulation of the algorithm in terms of excess quantities. To apply the algorithm to moving singularities, we should simply change the non-equilibrium jump relation for true inhomogeneities to another non-equilibrium jump relation valid for quasi-inhomogeneities. The details of the procedure is described in the next chapters.

Chapter 6

Macroscopic Dynamics of Phase-Transition Fronts

Martensitic phase-transition front propagation provides a good example of moving discontinuities.

As it is known, martensitic transformations are first order, diffusionless, shear solid state structural changes [Christian (1965)]. A cooperative rearrangement of atoms into a new, more stable crystal structure is observed in such a case of displacive transformation but without changing the chemical nature of the material. It follows that the austenite and martensite lattices will be intimately related. Incoherent interfaces are not compatible with these requirements and can be ruled out for martensitic transformations. Martensite plates can grow at speeds which approach that of sound in the material. Such large speeds are inconsistent with diffusion during transformation.

The propagation of phase interfaces in shape-memory alloys under applied stress is an experimentally observed phenomenon [Shaw and Kyriakides (1997); Sun, Li and Tse (2000)]. This phenomenon may be considered at different levels of description [Maugin (1998-1999)]. In the context of this book, we focus on macroscopic aspects of the dynamics of diffusionless stress-induced martensitic phase-transition fronts which are viewed as moving discontinuities in thermoelastic solids. From the mathematical point of view, such a problem is related to non-classical shock problems for conservation laws [Lefloch (2002); Dafermos (2005)].

The phase-transition front is represented by a jump discontinuity surface of zero thickness separating the different homogeneous austenite and martensite phases. This means that internal structure of the front [Truskinovsky (1987); Ngan and Truskinovsky (2002)] is beyond this framework.

The simplest formulation of the stress-induced phase-transition front propagation problem is given [Abeyaratne, Bhattacharya and Knowles

(2001)] in the case of an isothermal uniaxial motion of a slab. The phase front is represented by a jump discontinuity surface separating the different austenite and martensite branches of the N-shaped stress-strain curve. A shift of the martensitic branch of the curve is provided by the incorporation of a transformation strain, which is here considered as an experimentally determined material constant.

Standard boundary value problems do not have a unique solution when phase boundaries are present. The uniqueness of the solution is provided by the introduction of two supplementary constitutive-like relationships: a kinetic law for a driving force that establishes the speed of the transformation front and a nucleation criterion [Abeyaratne and Knowles (1990, 1991)].

A similar problem was analyzed on the basis of a constitutive model of the shape memory effect and pseudoelasticity of polycrystalline shape-memory alloy [Chen and Lagoudas (2000); Bekker et al. (2002); Shaw (2002); Lagoudas et al. (2003)]. The key feature of this approach is to introduce one or more internal variables (order parameters) describing the internal structure of the material. In the macroscopic modeling of shape-memory alloys, each point of the material represents a phase mixture. Local quantities are volume-averaged macroscopic quantities that depend on the microstructure through some overall descriptor (usually the phase fraction). Solution of the problem is provided by prescribing two constitutive ingredients: the macroscopic free energy function and a set of kinetic rate equations for the microstructural descriptors. The free energy function is decomposed in elastic and inelastic parts [Helm and Haupt (2003)]. The inelastic part represents the energy storage due to internal stress fields (internal variables). The resulting equations for the macroscopic behavior fit into the framework of internal variable models [Bernardini (2001)]. Several models fitting into this basic framework have been proposed although sometimes employing quite different formalisms [Fischer et al. (1994); Birman (1997); Bernardini and Pence (2002)]. All of them involve a constitutive information prescribed via state equations and kinetic equations for the internal variables. Differences between the models involve the choice and interpretation of the internal variables and the form of kinetic equations.

In spite of the differences between the above mentioned approaches, one assumption is common: both of them exploit the local equilibrium approximation though jump relations at the phase boundary may differ from the classical equilibrium jump relations. This means that all the fields including temperature and entropy are supposed to be well defined at any point as usual [Leo, Shield and Bruno (1993); Kim and Abeyaratne (1995); Shaw and Kyriakides (1997)]. At the same time, it is well understood that the martensitic phase transformation is a dissipative process that involves entropy change. The standard local equilibrium approximation is not sufficient to describe such a behavior. As discussed in chapter 3, the local phase equilibrium conditions should be taken into account.

In what follows we consider the simplest possible one-dimensional setting of the problem of impact-induced phase transformation front propagation. Both martensitic and austenitic phases are considered as isotropic materials. Since thermal expansion coefficient, for example, of NiTi is around $6 - 11 \times 10^{-6} \, \mathrm{K}^{-1}$, the thermal strain in the material is negligible under the variation up to 100 K. Therefore, the isothermal case is considered first. Then the adiabatic case is analyzed.

6.1   Isothermal impact-induced front propagation

The only experimental investigation concerning impact-induced austenite-martensite phase transformations was reported by Escobar and Clifton (1993, 1995). In their experiments, Escobar and Clifton used thin plate-like specimens of Cu-14.44Al-4.19Ni shape-memory alloy single crystal. One face of this austenitic specimen was subjected to an oblique impact loading, generating both shear and compression (Fig. 6.1). The conditions of the experiment were carefully designed so as to lead to plane wave propagation in the direction of the specimen surface normal. The orientation of the specimen relative to the lattice was chosen so as to activate only one single variant of martensite. The temperature changes during Escobar and Clifton's experiments are thought to be relatively unimportant. The measurements are taken in the central part of the rear face of the specimen. The ratio of the lateral to the transversal dimensions was chosen so that all the measurements were completed before the arrival of any release wave originating at the lateral faces of the slab. As Escobar and Clifton noted, measured velocity profiles provide several indications of the existence of a propagating phase boundary, in particular, a difference between the measured particle velocity and the transverse component of the projectile velocity. This velocity difference, in the absence of any evidence of plastic deformation, is indicative of a stress-induced phase transformation that propagates into the crystals from the impact face.

The determination of this velocity difference is most difficult from the

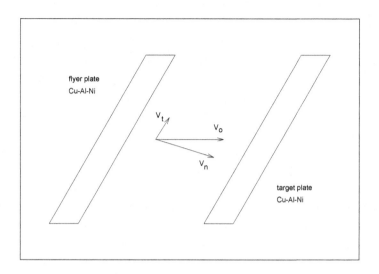

Fig. 6.1   Geometry of the experiment by Escobar and Clifton (1993, 1995).

theoretical point of view, because it depends on the velocity of the moving phase boundary.

A kinetic relation between the phase boundary velocity and the driving force was determined by Abeyaratne and Knowles (1997b) on the basis of the constitutive model they developed and the experimental data from Escobar and Clifton (1993, 1995). The resulting square-root kinetic curve is unbounded. This is in contradiction to their own result of the asymptotic behavior of kinetic curves [Abeyaratne and Knowles (2006)]. Therefore, the theoretical explanation of the particle velocity difference is still required.

6.1.1   *Uniaxial motion of a slab*

In order to model the conditions of the experiment by Escobar and Clifton (1993, 1995), it is sufficient to consider the problem in the one-dimensional setting.

Following Abeyaratne, Bhattacharya and Knowles (2001), we consider the simplest possible formulation of the problem, namely, the uniaxial motion of a slab. Consider a slab which, in an unstressed reference configuration occupies the region $0 < x_1 < L$, $-\infty < x_2, x_3 < \infty$, and consider the uniaxial motion of the form

$$u_1 = u(x,t), \quad x = x_1, \quad u_2 = u_3 \equiv 0, \tag{6.1}$$

where u_i are components of the displacement vector. In this case, we have only one non-vanishing component of the strain tensor

$$\varepsilon = \frac{\partial u}{\partial x}, \tag{6.2}$$

and, therefore, the bulk equations (4.7), (4.8) in the isothermal case are reduced to

$$\rho_0(x)\frac{\partial v}{\partial t} = \frac{\partial \sigma}{\partial x}, \tag{6.3}$$

$$\frac{\partial \sigma}{\partial t} = (\lambda(x) + 2\mu(x))\frac{\partial v}{\partial x}, \tag{6.4}$$

where v is the particle velocity

$$v = \frac{\partial u}{\partial t}, \tag{6.5}$$

and σ is the uniaxial stress.

The velocity and stress fields $v(x,t)$ and $\sigma(x,t)$ subject to the following initial and boundary conditions:

$$\sigma(x,0) = v(x,0) = 0, \quad \text{for} \quad 0 < x < L, \tag{6.6}$$

$$v(0,t) = v_0(t), \quad \sigma(L,t) = 0, \quad \text{for} \quad t > 0, \tag{6.7}$$

where $v_0(t)$ is a given time-dependent function, and to the jump relations at the moving phase boundary $\mathcal{S}$

$$\bar{V}_N[\rho_0 v] + [\sigma] = 0, \tag{6.8}$$

$$[V] = 0, \quad [\theta] = 0, \tag{6.9}$$

$$f_{\mathcal{S}} = -[W] + \langle \sigma \rangle [\varepsilon], \tag{6.10}$$

$$f_{\mathcal{S}}\bar{V}_N = \theta_{\mathcal{S}}\sigma_{\mathcal{S}} \geq 0. \tag{6.11}$$

Here V is the material velocity, $\bar{V}_N$ is the velocity of the moving phase boundary, W is the free energy per unit volume, which will be specified later.

The formulated uniaxial dynamic problem (6.3)-(6.11) is solved by means of the finite volume numerical scheme (4.56), (4.57) which can be represented in the considered case as follows:

$$(\bar{v})_n^{k+1} - (\bar{v})_n^k = \frac{\Delta t}{\Delta x} \frac{1}{\rho_n} \left((\Sigma^+)_n^k - (\Sigma^-)_n^k \right), \tag{6.12}$$

$$(\bar{\sigma})_n^{k+1} - (\bar{\sigma})_n^k = \frac{\Delta t}{\Delta x} (\lambda_n + 2\mu_n) \left((\mathcal{V}^+)_n^k - (\mathcal{V}^-)_n^k \right). \tag{6.13}$$

Remember that the superscript k denotes time step and subscript n corresponds to the computational cell location in space. The excess quantities Σ and $\mathcal{V}$ will be determined by means of the local equilibrium jump relations (3.47), (3.48) in the next sections.

6.1.2   *Excess quantities in the bulk*

The values of the excess quantities in bulk are determined in the same way as in the case of wave propagation. In the isothermal case, the nonequilibrium jump relation (3.47) reduces to

$$[\bar{\sigma} + \Sigma] = 0. \tag{6.14}$$

This means that at the interface between cells $(n-1)$ and (n) in the uniaxial case

$$(\Sigma^+)_{n-1} - (\Sigma^-)_n = (\bar{\sigma})_n - (\bar{\sigma})_{n-1}. \tag{6.15}$$

The jump relation following from the kinematic compatibility (4.22) reads

$$[\bar{v} + \mathcal{V}] = 0. \tag{6.16}$$

Therefore, we obtain in the uniaxial case

$$(\mathcal{V}^+)_{n-1} - (\mathcal{V}^-)_n = (\bar{v})_n - (\bar{v})_{n-1}. \tag{6.17}$$

Using the conservation of "Riemann invariants" for excess quantities (4.66), (4.67)

$$\left(\Sigma^- \right)_n + \rho_n c_n \left(\mathcal{V}^- \right)_n = 0, \tag{6.18}$$

$$\left(\Sigma^+ \right)_{n-1} - \rho_{n-1} c_{n-1} \left(\mathcal{V}^+ \right)_{n-1} = 0, \tag{6.19}$$

we obtain then the system of linear equations for the determination of excess quantities. This system of equations can be solved exactly for each boundary between computational cells. Then the field quantities can be updated for the next time step by means of the numerical scheme (6.12)-(6.13).

6.1.3 Excess quantities at the phase boundary

To determine the values of excess stresses at the moving phase boundary, we keep the continuity of excess stresses across the phase boundary [Berezovski and Maugin (2005b)]

$$[\Sigma] = 0, \qquad (6.20)$$

which yields

$$\left(\Sigma^+\right)_{p-1} - \left(\Sigma^-\right)_p = 0, \qquad (6.21)$$

where the phase boundary is placed between elements $(p-1)$ and (p).

The last jump condition can be interpreted as the *conservation of the genuine unknown jump at the phase boundary* in the numerical calculations because (6.20) means that

$$[\sigma] = [\bar{\sigma} + \Sigma] = [\bar{\sigma}]. \qquad (6.22)$$

This is illustrated in Fig. 6.2.

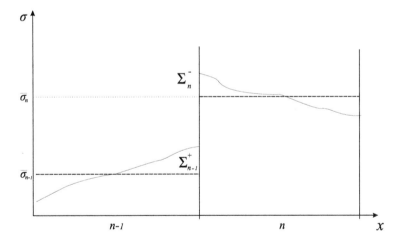

Fig. 6.2   Stresses at the moving boundary.

From the thermodynamical point of view, the continuity of excess stresses (6.20) means that we keep the same distance from equilibrium for both phases in our local equilibrium description (cf. Section 4.6).

The relation (6.20) should be complemented by the coherency condition between phases [Maugin (1998)]

$$[\mathbf{V}] = 0. \qquad (6.23)$$

The jump of the kinematic condition between material and physical velocity [Maugin (1993)] reads

$$[\mathbf{v} + \mathbf{F} \cdot \mathbf{V}] = [\mathbf{v}] + \langle \mathbf{F} \cdot \rangle [\mathbf{V}] + [\mathbf{F}] \cdot \langle \mathbf{V} \rangle = 0. \tag{6.24}$$

This means that in the uniaxial case

$$[v] + [\varepsilon]\langle V \rangle = 0. \tag{6.25}$$

Inserting the last relation into the jump relation following from the kinematic compatibility condition (5.2), we obtain the identity

$$\bar{V}_N [\varepsilon] - [\varepsilon]\langle V \rangle \equiv 0, \tag{6.26}$$

since the material velocity is continuous across the phase boundary.

To be consistent, we require the conservation of the genuine jump also for velocity

$$[\mathcal{V}] = 0, \tag{6.27}$$

which leads to the jump relations for linear momentum in terms of averaged quantities

$$\bar{V}_N [\rho_0 \bar{v}] + [\bar{\sigma}] = 0, \tag{6.28}$$

and to the corresponding jump of the kinematic compatibility

$$\bar{V}_N [\bar{\varepsilon}] + [\bar{v}] = 0. \tag{6.29}$$

We still keep the relations between excess stresses and excess velocities (6.18), (6.19). This means that in terms of excess stresses Eq. (6.27) yields

$$\frac{(\Sigma^+)_{p-1}}{\rho_{p-1} c_{p-1}} + \frac{(\Sigma^-)_p}{\rho_p c_p} = 0. \tag{6.30}$$

It follows from the conditions (6.21) and (6.30) that the values of excess stresses vanish at the phase boundary

$$(\Sigma^+)_{p-1} = (\Sigma^-)_p = 0. \tag{6.31}$$

Now all the excess quantities at the phase boundary are determined, and we can update the state of the elements adjacent to the phase boundary.

However, the proposed procedure should be applied at the phase boundary only after the initiation of the phase transition process. The possible motion of the interface between phases should also be taken into account.

6.1.4 *Initiation criterion for the stress-induced phase transformation*

Up to now we have used different local equilibrium jump relations in the bulk and at the phase boundary. It is expected that, at the beginning of the phase transformation process, the local equilibrium jump relations should be changed from one form to another. This means that at the initiation of the stress-induced phase transition both local equilibrium jump relations (3.47) and (3.48) are fulfilled at the phase boundary simultaneously:

$$\left[\bar{\theta} \left(\frac{\partial S}{\partial \varepsilon} \right)_\sigma + \bar{\sigma} + \Sigma \right] = 0, \tag{6.32}$$

$$\left[\bar{\sigma} - \bar{\theta} \left(\frac{\partial \bar{\sigma}}{\partial \theta} \right)_\varepsilon + \Sigma - \bar{\theta} \left(\frac{\partial \Sigma}{\partial \theta} \right)_\varepsilon \right] = 0, \tag{6.33}$$

where $S = \bar{S} + S_{ex}$ due to the additivity of entropy.

Eliminating the jumps of stresses from the system of Eqs. (6.32), (6.33), we obtain, then, a combined jump relation

$$\left[\bar{\theta} \left(\frac{\partial S}{\partial \varepsilon} \right)_\sigma + \bar{\theta} \left(\frac{\partial \bar{\sigma}}{\partial \theta} \right)_\varepsilon + \bar{\theta} \left(\frac{\partial \Sigma}{\partial \theta} \right)_\varepsilon \right] = 0. \tag{6.34}$$

We suppose that the temperature dependence for the excess stresses is the same as for averaged stresses

$$\left(\frac{\partial \bar{\sigma}}{\partial \theta} \right)_\varepsilon = \left(\frac{\partial \Sigma}{\partial \theta} \right)_\varepsilon = m, \tag{6.35}$$

where m is the thermoelastic coupling coefficient. Therefore, the combined non-equilibrium jump relation (6.34) takes the form

$$\left[\bar{\theta} \left(\frac{\partial S}{\partial \varepsilon} \right)_\sigma + 2 \bar{\theta} m \right] = 0. \tag{6.36}$$

Remembering the continuity of excess stresses at the phase boundary we obtain from Eq. (6.32)

$$[\bar{\sigma}] = - \left[\bar{\theta} \left(\frac{\partial S}{\partial \varepsilon} \right)_\sigma \right]. \tag{6.37}$$

Therefore, we will finally have

$$[\bar{\sigma}] = 2\theta_0 [m]. \tag{6.38}$$

This is the desired criterion of the initiation of phase transformation in terms of the jump of the averaged stresses.

It is more convenient to express the obtained initiation criterion in terms of the driving force using Eq. (6.37)

$$\left[\bar{\theta} \left(\frac{\partial S}{\partial \varepsilon} \right)_\sigma \right] = -2\theta_0 \, [m] \, . \tag{6.39}$$

To be able to calculate the jump of the entropy derivative, we suppose that entropy behaves like the driving force in the neighborhood of the phase boundary (see Sec. 3.5)

$$S = \frac{f}{\theta}, \quad f_{\mathcal{S}} = [f], \tag{6.40}$$

where f is defined by analogy with Eq. (6.10)

$$f = -\bar{W} + \langle \bar{\sigma} \rangle \bar{\varepsilon}, \tag{6.41}$$

with

$$\langle \bar{\sigma} \rangle = \frac{(\bar{\sigma})_p + (\bar{\sigma})_{p-1}}{2}. \tag{6.42}$$

This supposes that, at the point where Eqs. (6.40) and (6.41) are defined, there exists in thought an oriented surface of unit normal $\mathbf{n}$. If there is no discontinuity across this surface, then f is a so-called generating function (the complementary energy changed of sign and up to a constant). If there does exist a discontinuity, then the expression becomes meaningful only if the operator $[\cdots]$ is applied to it.

Using the representation (6.40), we obtain for the derivative of the entropy

$$\left(\frac{\partial S}{\partial \varepsilon} \right)_\sigma = \frac{1}{\theta} \left(\frac{\partial f}{\partial \varepsilon} \right)_\sigma - \frac{f}{\theta^2} \left(\frac{\partial \theta}{\partial \varepsilon} \right)_\sigma . \tag{6.43}$$

Substituting the last relation into Eq. (6.37), we can determine the stress jump at the phase boundary in terms of the driving force

$$[\bar{\sigma}] = f_{\mathcal{S}} \left\langle \frac{1}{\bar{\theta}} \left(\frac{\partial \bar{\theta}}{\partial \varepsilon} \right)_\sigma \right\rangle + \langle f \rangle \left[\frac{1}{\bar{\theta}} \left(\frac{\partial \bar{\theta}}{\partial \varepsilon} \right)_\sigma \right] - \left[\left(\frac{\partial f}{\partial \varepsilon} \right)_\sigma \right] . \tag{6.44}$$

This means that we can specify the combined jump relation at the phase boundary (6.38) to the form

$$f_{\mathcal{S}} \left\langle \frac{1}{\bar{\theta}} \left(\frac{\partial \bar{\theta}}{\partial \varepsilon} \right)_\sigma \right\rangle + \langle f \rangle \left[\frac{1}{\bar{\theta}} \left(\frac{\partial \bar{\theta}}{\partial \varepsilon} \right)_\sigma \right] - \left[\left(\frac{\partial f}{\partial \varepsilon} \right)_\sigma \right] = 2\theta_0 [m]. \tag{6.45}$$

In the uniaxial case, the expression for the function f in the transformed martensitic part can be written as follows

$$f = -\frac{1}{2}\bar{\sigma}(\bar{\varepsilon} - \varepsilon_{tr}) + \frac{1}{2}\frac{C}{\theta_0}(\bar{\theta} - \theta_0)^2 + \sigma_{tr}\varepsilon_{tr}$$
$$-\frac{m}{2}(\bar{\theta} - \theta_0)(\bar{\varepsilon} - \varepsilon_{tr}) + \langle\bar{\sigma}\rangle(\bar{\varepsilon} - \varepsilon_{tr}) + f_0,$$

(6.46)

where ε_{tr} is the transformation strain, σ_{tr} is the transformation stress, f_0 is an arbitrary constant. Correspondingly, in the austenitic phase

$$f = -\frac{1}{2}\bar{\sigma}\bar{\varepsilon} + \frac{1}{2}\frac{C}{\theta_0}(\bar{\theta} - \theta_0)^2 - \frac{m}{2}(\bar{\theta} - \theta_0)\bar{\varepsilon} + \langle\bar{\sigma}\rangle\bar{\varepsilon} + f_0,$$

(6.47)

where elastic coefficients are different for austenite and martensite, but the constant f_0 is the same. Therefore, the derivative of f with respect to the strain at fixed stress in both cases is given by

$$\left(\frac{\partial f}{\partial \varepsilon}\right)_\sigma = -\frac{\bar{\sigma}}{2} - \frac{m}{2}(\bar{\theta} - \theta_0) + \langle\bar{\sigma}\rangle,$$

(6.48)

because the temperature is determined independently from other fields in the thermal stress approximation. The corresponding jump in the isothermal case is

$$\left[\left(\frac{\partial f}{\partial \varepsilon}\right)_\sigma\right] = -\frac{1}{2}[\bar{\sigma}].$$

(6.49)

Substituting the obtained value of the jump into Eq. (6.45) and taking into account Eq. (6.38), we have

$$f_s\left\langle\frac{1}{\bar{\theta}}\left(\frac{\partial\bar{\theta}}{\partial\varepsilon}\right)_\sigma\right\rangle + \langle f\rangle\left[\frac{1}{\bar{\theta}}\left(\frac{\partial\bar{\theta}}{\partial\varepsilon}\right)_\sigma\right] = \theta_0[m],$$

(6.50)

or, more explicitly,

$$f_s\left\langle\frac{(\lambda + 2\mu)}{m}\right\rangle + \langle f\rangle\left[\frac{(\lambda + 2\mu)}{m}\right] = -\theta_0^2[m].$$

(6.51)

In the correspondence with Eqs. (6.46) and (6.47), the mean value of the function f is expressed in the isothermal case as

$$\langle f\rangle = \frac{1}{4}\left(\bar{\sigma}_M\bar{\varepsilon}_A + \bar{\sigma}_A(\bar{\varepsilon}_M - \varepsilon_{tr}) + 2\sigma_{tr}\varepsilon_{tr}\right) + f_0,$$

(6.52)

whereas its jump is

$$[f] = \frac{1}{2}\left(\bar{\sigma}_M\bar{\varepsilon}_A - \bar{\sigma}_A(\bar{\varepsilon}_M - \varepsilon_{tr}) - 2\sigma_{tr}\varepsilon_{tr}\right),$$

(6.53)

where subscripts "A" and "M" denote austenite and martensite, respectively.

Usually, the transformation stress σ_{tr} and the transformation strain ε_{tr} are treated as material parameters [Abeyaratne, Bhattacharya and Knowles (2001)]. Therefore, we can rewrite the relations (6.52) and (6.53) as follows

$$\langle f \rangle - f_0 - \frac{1}{2}\sigma_{tr}\varepsilon_{tr} = \frac{\bar{\sigma}_A \bar{\sigma}_M}{2} \left\langle \frac{1}{\lambda + 2\mu} \right\rangle, \tag{6.54}$$

$$[f] + \sigma_{tr}\varepsilon_{tr} = \frac{\bar{\sigma}_A \bar{\sigma}_M}{2} \left[\frac{1}{\lambda + 2\mu} \right]. \tag{6.55}$$

Dividing both parts of Eq. (6.54) by the corresponding parts of Eq. (6.55) we can eliminate stresses from the relation between mean value and jump of the function f

$$\frac{\langle f \rangle - f_0 - \frac{1}{2}\sigma_{tr}\varepsilon_{tr}}{[f] + \sigma_{tr}\varepsilon_{tr}} = \left\langle \frac{1}{\lambda + 2\mu} \right\rangle \left[\frac{1}{\lambda + 2\mu} \right]^{-1}. \tag{6.56}$$

After rearranging the last equation, we will have for the mean value of the function f

$$\langle f \rangle = f_0 + \frac{1}{2}\sigma_{tr}\varepsilon_{tr} + ([f] + \sigma_{tr}\varepsilon_{tr}) \left\langle \frac{1}{\lambda + 2\mu} \right\rangle \left[\frac{1}{\lambda + 2\mu} \right]^{-1}. \tag{6.57}$$

Substituting the relation (6.57) into Eq. (6.51) and remembering that the jump of the function f is the driving force f_S, we obtain

$$\begin{aligned} f_S \left[\frac{1}{m} \right] &\left[\frac{1}{\lambda + 2\mu} \right]^{-1} = -\theta_0^2[m] \\ &- \left(f_0 + \frac{1}{2}\sigma_{tr}\varepsilon_{tr} + \sigma_{tr}\varepsilon_{tr} \left\langle \frac{1}{\lambda + 2\mu} \right\rangle \left[\frac{1}{\lambda + 2\mu} \right]^{-1} \right) \left[\frac{(\lambda + 2\mu)}{m} \right]. \end{aligned} \tag{6.58}$$

Therefore, the combined jump relation at the phase boundary (6.58) determines the value of the driving force at the interface. This value depends on the choice of the constant f_0. It is easy to see that the simplest expression can be obtained if we choose the value of f_0 as

$$f_0 = -\frac{1}{2}\sigma_{tr}\varepsilon_{tr} - \sigma_{tr}\varepsilon_{tr} \left\langle \frac{1}{\lambda + 2\mu} \right\rangle \left[\frac{1}{\lambda + 2\mu} \right]^{-1}. \tag{6.59}$$

In the isothermal case, we can assume even zero mean value of the function f in Eq. (6.51) [Berezovski and Maugin (2005b)], since at the beginning of the phase transformation, the transformation stress should be equal to the difference in stress in the martensitic and austenitic phases

$$\sigma_{tr} = \bar{\sigma}_M - \bar{\sigma}_A, \tag{6.60}$$

and the transformation strain is the strain difference between the transformed martensite and austenite

$$\varepsilon_{tr} = (\bar{\varepsilon}_M - \varepsilon_{tr}) - \bar{\varepsilon}_A. \tag{6.61}$$

However, this is not possible in the adiabatic case, as we will see below. Therefore, we keep the choice (6.59) which leads to the following expression for the driving force

$$f_S = -\theta_0^2 [m] \left[\frac{1}{m} \right]^{-1} \left[\frac{1}{\lambda + 2\mu} \right]. \tag{6.62}$$

The right-hand side of the latter relation can be interpreted as a critical value of the driving force in the uniaxial case. Therefore, the proposed criterion for the beginning of the isothermal stress-induced phase transition is the following one:

$$|f_S| \geq |f_{critical}|, \tag{6.63}$$

where

$$f_{critical} = \theta_0^2 [m] \left[\frac{1}{m} \right]^{-1} \left[\frac{1}{\lambda + 2\mu} \right]. \tag{6.64}$$

Note that we have eliminated the contribution of the transformation strain by means of the choice of the constant f_0. Moreover, the critical value of the driving force should be equal to zero if the properties of martensite and austenite are identical.

6.1.5 *Velocity of the phase boundary*

We have proposed the satisfaction of the non-equilibrium jump relation at the phase boundary

$$\left[\bar{\theta} \left(\frac{\partial S}{\partial \varepsilon} \right)_\sigma + \bar{\sigma} + \Sigma \right] = 0. \tag{6.65}$$

Moreover, we have assumed the continuity of the excess stresses at the phase boundary (6.20). This means that the non-equilibrium jump relation (6.65) is simplified to

$$\left[\bar{\theta} \left(\frac{\partial S}{\partial \varepsilon} \right)_\sigma + \bar{\sigma} \right] = 0. \tag{6.66}$$

Substituting the results (6.43), (6.44), (6.49), and (6.59) into Eq. (6.66) we can express the stress jump at the phase boundary in terms of the driving force

$$f_S \left[\frac{1}{m}\right] = -\frac{\theta_0[\bar{\sigma}]}{2}\left[\frac{1}{\lambda + 2\mu}\right]. \tag{6.67}$$

Having the value of the stress jump, we can determine the material velocity at the moving discontinuity by means of the jump relation for linear momentum (6.28)

$$\bar{V}_N[\rho_0\bar{v}] + [\bar{\sigma}] = 0. \tag{6.68}$$

The jump of kinematic compatibility (6.29) gives

$$[\bar{v}] = -[\bar{\varepsilon}]\bar{V}_N, \tag{6.69}$$

and the jump relation for linear momentum (6.68) can be rewritten in a form that is more convenient for the calculation of the velocity at the phase boundary

$$\rho_0\bar{V}_N^2[\bar{\varepsilon}] = [\bar{\sigma}]. \tag{6.70}$$

In the isothermal case we have for the strain jump

$$[\bar{\varepsilon}] = \left[\frac{\bar{\sigma}}{\lambda + 2\mu}\right] - \varepsilon_{tr}, \tag{6.71}$$

or

$$[\bar{\varepsilon}] = [\bar{\sigma}]\left\langle\frac{1}{\lambda + 2\mu}\right\rangle + \langle\bar{\sigma}\rangle\left[\frac{1}{\lambda + 2\mu}\right] - \varepsilon_{tr}. \tag{6.72}$$

Representing the mean value of the stress as

$$\langle\bar{\sigma}\rangle = \sigma_A - [\bar{\sigma}]/2, \tag{6.73}$$

we can rewrite the expression for the stress jump at the phase boundary as follows

$$[\bar{\varepsilon}] = A[\bar{\sigma}] - B, \tag{6.74}$$

where

$$A = \left\langle\frac{1}{\lambda + 2\mu}\right\rangle - \frac{1}{2}\left[\frac{1}{\lambda + 2\mu}\right], \quad B = \varepsilon_{tr} - \sigma_A\left[\frac{1}{\lambda + 2\mu}\right]. \tag{6.75}$$

Therefore, Eq. (6.70) can be represented in terms of the stress jump

$$\rho_0\bar{V}_N^2 = \frac{[\bar{\sigma}]}{A[\bar{\sigma}] - B}, \tag{6.76}$$

or, in terms of the driving force

$$\rho_0 \bar{V}_N^2 = \frac{f_S}{A f_S - B/D}, \tag{6.77}$$

where, in the correspondence with Eq. (6.67)

$$D = -2 \left[\frac{1}{\theta_0 m} \right] \left[\frac{1}{\lambda + 2\mu} \right]^{-1}. \tag{6.78}$$

Introducing a "characteristic" velocity c_* by

$$\frac{1}{c_*^2} = A\rho_0 = \rho_0 \left(\left\langle \frac{1}{\lambda + 2\mu} \right\rangle - \frac{1}{2} \left[\frac{1}{\lambda + 2\mu} \right] \right), \tag{6.79}$$

we have, finally

$$\bar{V}_N^2 / c_*^2 = \frac{f_S}{f_S - B/AD}. \tag{6.80}$$

Another form of the kinetic relation can be obtained by inverting Eq. (6.80)

$$f_S = \frac{B}{AD} \left(1 - \frac{1}{1 - \bar{V}_N^2 / c_*^2} \right). \tag{6.81}$$

It is instructive to compare the obtained kinetic relation with that in Purohit and Bhattacharya (2003)

$$f_S = E\varepsilon_{tr} \left(\varepsilon_0 - \frac{1 + \varepsilon_{tr}/2}{1 - \bar{V}_N^2 / c_*^2} \right). \tag{6.82}$$

As one can see, the functional form of the kinetic relations is the same.

Thus, the kinetic relation and the initiation criterion are derived from the local equilibrium jump relations (6.32), (6.33) under assumption of the continuity of excess quantities at the phase boundary in the uniaxial isothermal case.

It should be noted that the phase transformation process begins only if the value of the driving force is over the critical value (6.64). Therefore, we exploit $f_S - f_{critical}$ instead of f_S in the expression (6.77). However, the strain jump is still calculated by f_S. This means that the final expression for the kinetic relation has the form

$$\rho_0 \bar{V}_N^2 = \frac{D(f_S - f_{critical})}{AD(f_S - f_{critical}) + ADf_{critical} - B}. \tag{6.83}$$

The obtained relations at the phase boundary are used in the numerical scheme described above for the simulation of phase-transition front propagation.

6.2   Numerical simulations

6.2.1   *Algorithm description*

As it was noted, the excess quantities correspond to the cell-centered numerical fluxes for the conservative wave-propagation algorithm [Bale et al. (2003)]. This means that we can exploit the advantages of the wave-propagation algorithm [LeVeque (1997)]. However, no limiters are used in the calculations. Suppressing spurious oscillations is achieved by means of using a first-order Godunov step after each three second-order Lax-Wendroff steps. This idea of composition was invented by Liska and Wendroff (1998).

A procedure similar to a cellular automaton is applied to the phase-transition front tracking. At any time step, the values of the driving force are calculated in cells adjacent to the phase boundary. If the value of the driving force at the phase boundary exceeds the critical one, the velocity of the phase front is computed by means of Eq. (6.70). Virtual displacements of the phase-transition front are calculated then for all possible phase boundaries adjacent to the cell. We keep the cell in the old phase state if the (algebraic) sum of the virtual displacements is less than the size of the space step, and change it to another phase otherwise.

All the calculations are performed with Courant-Friedrichs-Levy number equal to 1. In the homogeneous case this gives the exact solution of the hyperbolic system of Eqs. (6.3) and (6.4).

6.2.2   *Comparison with experimental data*

As the first result, the time-history for the particle velocity is given in Fig. 6.3. The time-history of the particle velocity shows that we have nothing early on because no waves have reached the considered point. After the coming of a fastest elastic wave, the jump in the particle velocity is observed, but its amplitude is lower than the initial one. This amplitude is restored in the second jump, which is associated with a phase transition front. Similar results were qualitatively predicted by Abeyaratne and Knowles (1997a) and obtained numerically for test problems by Zhong, Hou and LeFloch (1996).

To compare the results of modeling with experimental data by Escobar and Clifton (1993, 1995), the calculations of the particle velocity were performed for different impact velocities. Following Abeyaratne and Knowles

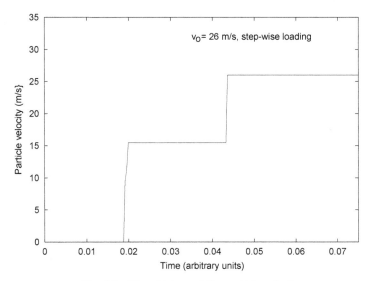

Fig. 6.3   Time-history of the particle velocity.

(1997b), we use the half velocity of the flyer plate because the specimen and the flyer plate were both composed of the Cu-Al-Ni alloy being tested. The half of the measured transverse particle velocity is chosen for the comparison due to reflection at the rear face of the specimen. The comparison of calculations with experimental data is presented in Fig. 6.4. As a result, we can see that the computed particle velocity is practically independent of the impact velocity. This constant value of the particle velocity corresponds to the constant shear stress at the phase boundary (cf. Fig. 6.4).

The kinetics of transformation can be represented by the relation between phase boundary speed and driving force, which is shown in Fig. 6.5. Here the corresponding values of the phase boundary speed are calculated by means of the relation (6.70), while the values of driving force are determined by Eq. (6.10). As previously, calculations were performed for different impact velocities. As one can see, the shape of the obtained curve is a nonlinear one. This confirms that linear kinetic relations are unable to predict the experimentally observed difference between tangential impact velocity and transversal particle velocity in the experiments by Escobar and Clifton (1993, 1995).

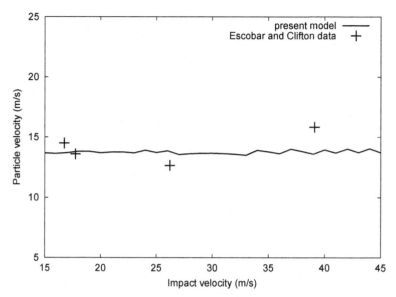

Fig. 6.4   Particle velocity versus impact velocity.

6.3   Interaction of a plane wave with phase boundary

Materials capable of phase transformation display a behavior known as superelasticity or pseudoelasticity [Christian (1965)]. Pseudoelasticity refers to the ability of the material to accommodate strains during loading and then recover upon unloading (via a hysteresis loop). In order to analyze the strain-stress behavior in the model described above, we consider the interaction of a plane wave with phase boundary.

The geometry of the problem is shown in Fig. 6.6. The wave is excited at the left boundary of the computation domain by prescribing a time variation of a component of the stress tensor. Top and bottom boundaries are stress-free, the right boundary is assumed to be rigid. The time-history of loading is shown in Fig. 6.7. If the magnitude of the wave is high enough, the phase transformation process is activated at the phase boundary. The maximal value of the Gaussian pulse is chosen as 0.7 GPa. The considered situation is similar to that in experiments of Escobar and Clifton (1993), where a plane phase transition front propagated in the normal direction to the plane of loading, and only one variant of martensite was activated.

Material properties correspond to Cu-14.44Al-4.19Ni shape-memory alloy [Escobar and Clifton (1993)] in austenitic phase: the density $\rho = 7100$

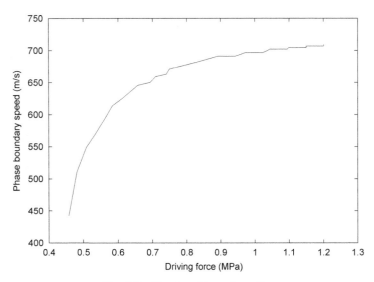

Fig. 6.5   Kinetics of transformation.

kg/m^3, the elastic modulus E = 120 GPa, the shear wave velocity c_s = 1187 m/s, the dilatation coefficient α = 6.75 · 10^{-6} 1/K. It was recently reported [Emel'yanov et al. (2000)] that elastic properties of martensitic phase of Cu-Al-Ni shape-memory alloy after impact loading are very sensitive to the amplitude of loading. Therefore, for the martensitic phase we choose, respectively, E = 60 GPa, c_s = 1055 m/s, with the same density and dilatation coefficient as above. As a first result of computations, the stress-strain relation is plotted in Fig. 6.8 at a fixed point inside the computational domain which was initially in the austenitic state.

As we can see, the stress-strain relation is at first linear, corresponding to elastic austenite. Then the strain value jumps along a constant stress line to its value in the martensitic state due to the phase transformation. Accordingly, both loading and unloading correspond to elastic martensite. The value of the strain jump between straight lines, the slope of which is prescribed by material properties of austenite and martensite, respectively, is determined by the value of stress that conforms to the critical value of the driving force (Eq. (6.54)). The critical value of the driving force should agree with the barrier of potential that we have to overcome to go from one phase to the other. Therefore, the stress value corresponding to the critical value of the driving force can be associated with the transformation stress, and the value of the strain jump is nothing else but the transformation

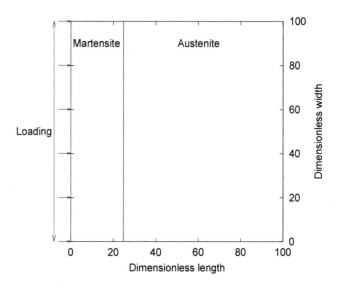

Fig. 6.6   Plane wave: geometry.

strain. The stress-strain behavior shown in Fig. 6.8 looks very much like the stress-strain dependence discussed in [Abeyaratne, Bhattacharya and Knowles (2001)].

This can also be represented in terms of free energy (Fig. 6.9), which has a typical two energy-wells structure like it is assumed in [Abeyaratne, Bhattacharya and Knowles (2001)].

6.3.1 Pseudoelastic behavior

Until now it was supposed that austenite is not recovered after unloading. If the value of the reference temperature is above the onset of reverse transformation temperature, we should expect that the austenitic state will be recovered after unloading. The inverse phase transformation should occur immediately when the actual deformation of martensitic elements becomes less than the transformation strain. Since the inverse transformation is governed by another condition than the direct transformation, we obtain a hysteretic stress-strain behavior (Fig. 6.10).

The obtained stress-strain relation at any fixed point results in an overall pseudoelastic response of a specimen.

To validate the obtained results, the local stress-strain relation at a fixed point which was initially in austenitic state was calculated. At given

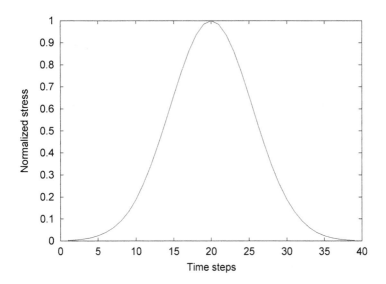

Fig. 6.7   Loading time-history.

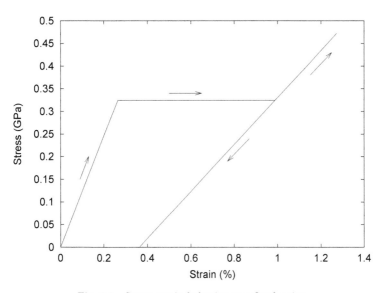

Fig. 6.8   Stress-strain behavior at a fixed point.

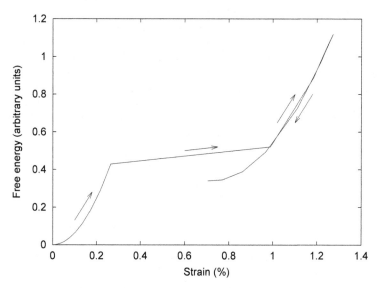

Fig. 6.9   Free energy at a fixed point.

temperature, martensite can exist only in the deformed state. Therefore, the transformation strain should be added in the martensitic state. The best fitting of experimental observations by McKelvey and Ritchie (2000) corresponds to the value for the transformation strain 3.3%. Just this value is used for the comparison with the experimental data. The calculated local stress-strain relation is plotted in Fig. 6.11 together with experimental data for the quasi-static loading from [McKelvey and Ritchie (2000)].

As we can see, the local stress-strain behavior predicted by the developed model is very similar to the experimental stress-strain curve for the bar under quasi-static loading. In quasi-statics, all the material points are deformed similarly. This means that the global behavior observed in experiments should be the same as the local one. Therefore, we can conclude that the proposed model describes well the observed behavior of the stress-strain curve.

The obtained stress-strain behavior results in a specific interaction between a plane stress wave and the phase boundary. The result of the interaction is shown in Fig. 6.12 that represents the structure of the interaction of the incoming wave (from left) with the phase boundary. Here we return to the loading pulse shown in Fig. 6.7. We observe that the amplitudes of both transmitted and reflected waves are cut down, whereas the phase boundary has moved to the austenitic region from its initial position

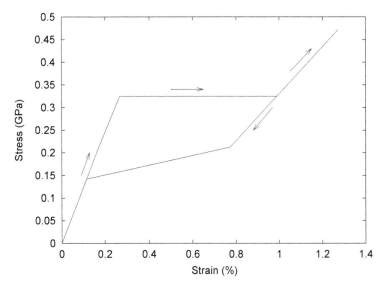

Fig. 6.10   Stress-strain behavior at a fixed point with full recovering of austenite.

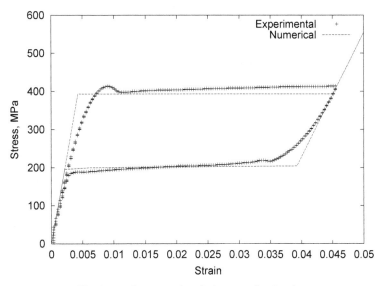

Fig. 6.11   Stress-strain relation at a fixed point.

shown in Fig. 6.6. Therefore, the martensitic phase transformation exhibits a property to be a filter for the amplitudes of incoming waves. The magnitude of the transmitted wave in Fig. 6.12 corresponds to the value of the

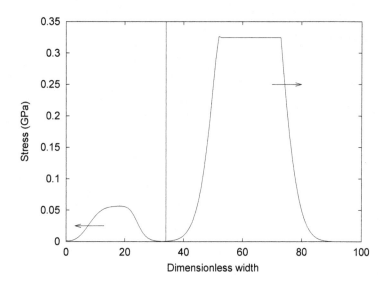

Fig. 6.12   Plane wave profile after interaction with moving phase boundary.

transformation stress (see Fig. 6.8).

The considered mathematical model of martensitic phase transition front propagation is as simple as possible. Martensite and austenite phases are treated as isotropic linear thermoelastic materials. The phase transition front is viewed as an ideal mathematical discontinuity surface. Only one variant of martensite is involved. This means that the phenomenon is highly idealized but one of the essential features of martensitic transformation holds: the considered phase transformation is diffusionless. The problem remains non-linear even in this simplified description that supports a numerical solution.

The supplementary constitutive information needed to avoid the non-uniqueness of the solution of the boundary-value problem is provided by means of local equilibrium jump relations at the moving phase boundary, which are formulated in terms of excess quantities. The same excess quantities are used in the construction of a finite volume numerical scheme that coincides with the conservative wave propagation algorithm in the absence of phase transformation. The continuity of the excess quantities at the phase boundary is assumed, which leads to the conservation of genuine jumps at the phase boundary in the finite volume algorithm. As a result, a closed system of governing equations and jump relations can be solved numerically. Results of numerical simulations show that the proposed approach allows us to capture experimental observations while corresponding to theoretical predictions in spite of the idealization of the process.

6.4 One-dimensional adiabatic fronts in a bar

The rise of temperature at a propagating martensitic phase-transition front was observed experimentally [Shaw and Kyriakides (1995, 1997); Orgeas and Favier (1998)], predicted theoretically [Abeyaratne and Knowles (1993, 1994a); Kim and Abeyaratne (1995); Maugin and Trimarco (1995)], and simulated numerically [Leo, Shield and Bruno (1993); Chen and Lagoudas (2000); Shaw (2000); Bekker et al. (2002); Chrysochoos, Licht and Peyroux (2003); Iadicola and Shaw (2004)].

As in the isothermal case, the determination of the velocity of the phase transition front is also an open question, since almost all simulations are performed under quasi-static conditions. The only exception is provided by papers by Lagoudas et al. [Chen and Lagoudas (2000); Bekker et al. (2002); Lagoudas et al. (2003)]. However, these authors determine the velocity of the front by means of the jump relation for linear momentum. As noted in [Maugin (1997)] this is possible only after having the solution for field variables in which the velocity of the front is intrinsically involved. That is why we extend the results for homothermal stress-induced phase transition front propagation [Berezovski and Maugin (2005b)] to the adiabatic case. The one-dimensional analysis of the phase front propagation in a bar is carried under the assumption that the change in cross-sectional area of the bar can be neglected.

Lagoudas et al. [Chen and Lagoudas (2000); Bekker et al. (2002); Lagoudas et al. (2003)] have clearly shown that, in an impact problem, the heat conduction is insignificant because of rapid loading. It is then a reasonable approximation to assume that the processes considered are adiabatic with heat conduction terms being ignored [Chen and Lagoudas (2000); Bekker et al. (2002); Lagoudas et al. (2003)]. In the adiabatic case, temperature is determined independently of other fields.

6.4.1 *Formulation of the problem*

We consider the boundary value problem of the tensile impact loading of a 1-D, shape-memory alloy bar that is initially in an austenitic phase and that has uniform cross-sectional area A_0. The bar occupies the interval $0 < x < L$ in a reference configuration and the boundary $x = 0$ is subjected to the tensile shock loading

$$\sigma(0, t) = \hat{\sigma} \quad \text{for} \quad t > 0. \tag{6.84}$$

The right end is kept traction free or fixed.

The bar is assumed to be long compared to its diameter so it is under uniaxial stress state and the stress $\sigma(x, t)$ depends only on the axial position and time. The temperature $\theta_0(t)$ of the environment surrounding the bar is prescribed, and the bar is initially at the same temperature as the environment.

The motion of the bar is characterized not only by the displacement field $u(x, t)$ as in the isothermal case, but also by the temperature field $\theta(x, t)$, where x denotes the location of a particle in the reference configuration and t is time. Linearized strain is further assumed so the axial component of the strain $\varepsilon(x, t)$ is related to the displacement by $\varepsilon = u_x(x, t)$. The density of the material ρ is assumed constant. All field variables are averaged over the cross-section of the bar.

We suppose that the bar is composed of a thermoelastic material characterized by its Helmholtz free energy per unit volume $W(\varepsilon, \theta)$. For such a material, the stress σ and entropy per unit volume S are determined by the constitutive relations

$$\sigma = \frac{\partial W}{\partial \varepsilon}, \quad S = -\frac{\partial W}{\partial \theta}. \tag{6.85}$$

The free energy per unit volume in linear isotropic inhomogeneous thermoelasticity is given by (see chapter 4)

$$W\left(\varepsilon, \theta; x\right) = \frac{1}{2}\left(\lambda(x) + 2\mu(x)\right)\varepsilon^2 - \frac{C(x)}{2\theta_0}\left(\theta - \theta_0\right)^2 + m(x)\left(\theta - \theta_0\right)\varepsilon, \tag{6.86}$$

where $C(x) = \rho_0 c$, c is the specific heat at constant stress, and only small deviations from a spatially uniform reference temperature θ_0 are envisaged. The *dilatation coefficient* α is related to the thermoelastic coefficient m, and the Lamé coefficients λ and μ by $m = -\alpha(3\lambda + 2\mu)$.

The constitutive relation (6.86) is valid for the austenitic phase. Martensite exists at the same temperature only in a transformed state which is characterized by a transformation strain ε_{tr}, a transformation stress σ_{tr}, and a latent heat L_{tr} [Abeyaratne and Knowles (1993)]

$$W\left(\varepsilon, \theta; x\right) = \frac{1}{2}\left(\lambda(x) + 2\mu(x)\right)\left(\varepsilon - \varepsilon_{tr}\right)^2 - \frac{C(x)}{2\theta_0}\left(\theta - \theta_0\right)^2$$
$$+ m(x)\left(\theta - \theta_0\right)\left(\varepsilon - \varepsilon_{tr}\right) - \sigma_{tr}\varepsilon_{tr} - \frac{L_{tr}}{\theta_0}(\theta - \theta_0). \tag{6.87}$$

In the thermomechanical process, the one-dimensional balance of linear momentum and kinematic compatibility, which are the same as in the isothermal case

$$\rho(x)\frac{\partial v}{\partial t} = \frac{\partial \sigma}{\partial x}, \tag{6.88}$$

$$\frac{\partial \sigma}{\partial t} = (\lambda(x) + 2\mu(x))\frac{\partial v}{\partial x}, \tag{6.89}$$

are complemented by the first and second laws of thermodynamics (cf. [Abeyaratne and Knowles (1994a)]) that read in the reference configuration

$$\theta\frac{\partial S}{\partial t} + \frac{\partial q}{\partial x} = r, \tag{6.90}$$

$$q\frac{\partial \theta}{\partial x} \leq 0, \tag{6.91}$$

where $v(x,t) = u_t$ is the particle velocity, $q(x,t)$ is the heat flux in the x-direction and $r(x,t)$ is the heat supply rate per unit volume.

Suppose that at time t there is a moving phase boundary at $x = \mathcal{S}(t)$. Then one also has the corresponding jump relations [Abeyaratne and Knowles (1994a)]

$$\rho\bar{V}_N[v] + [\sigma] = 0, \tag{6.92}$$

$$\bar{V}_N[\varepsilon] + [v] = 0, \tag{6.93}$$

$$\bar{V}_N\langle\theta\rangle[S] = [q] + f_{\mathcal{S}}\bar{V}_N, \tag{6.94}$$

where $\bar{V}_N$ is the material velocity of $\mathcal{S}$ and the driving traction $f_{\mathcal{S}}(t)$ at the discontinuity is defined by [Truskinovsky (1987); Abeyaratne and Knowles (1994a,b)]

$$f_{\mathcal{S}} = -[W] + \langle\sigma\rangle[\varepsilon] - \langle S\rangle[\theta]. \tag{6.95}$$

The second law of thermodynamics requires that $q\theta_x \leq 0$ where the fields are smooth, and that

$$f_{\mathcal{S}}\bar{V}_N \geq 0 \tag{6.96}$$

at the phase boundary. If $f_{\mathcal{S}}$ is not zero, the sign of $\bar{V}_N$, and hence the direction of motion of discontinuity, is determined by the sign of $f_{\mathcal{S}}$.

6.4.2   Adiabatic approximation

The time scale of the dynamic problem is of the order of micro- to milliseconds. The physically meaningful process is an adiabatic one because such time intervals are too short for heat conduction to take place as well as for convection to remove heat through the surface of the bar. In the adiabatic approximation, therefore, the heat conduction term q in Eq. (6.90) can be neglected [Lagoudas et al. (2003)].

Therefore, we can then rewrite the relevant *bulk* equations describing an adiabatic thermomechanical process in an inhomogeneous linear isotropic bar as the following three equations (cf. [Abeyaratne and Knowles (1994b)])

$$\rho(x)\frac{\partial v}{\partial t} = \frac{\partial \sigma}{\partial x}, \tag{6.97}$$

$$\frac{\partial \sigma}{\partial t} = (\lambda(x) + 2\mu(x))\frac{\partial v}{\partial x}, \tag{6.98}$$

$$\frac{\partial S}{\partial t} = 0. \tag{6.99}$$

Across a phase boundary, the jump relations (6.92)-(6.94) are still valid. Furthermore, if we compare the two heat sources (due to thermoelasticity and due to phase transformation), it becomes clear that the thermoelastic heat-generation term can be disregarded, and the temperature changes only due to the phase transformation [Lagoudas et al. (2003)]. This means that the jump relation (6.94) reduces in the adiabatic approximation to

$$\bar{V}_N\langle\theta\rangle[S] = f_S\bar{V}_N. \tag{6.100}$$

The indeterminacy of the velocity (and position) of the phase transition front leads to the non-linearity of the problem. As previously, we apply the same numerical method (6.12)-(6.13) for the solution of the formulated problem. The excess quantities at the interfaces between computational cells both in the bulk and at the phase boundary are determined by means of the local equilibrium jump relations, as it was described in section 6.2.

6.4.3   Initiation criterion for the stress-induced phase transformation in adiabatic case

In the adiabatic case, the initiation criterion for the stress-induced phase transformation should be modified since the jump of entropy is determined by the scalar value of the driving force (6.100) and the temperature at a discontinuity

$$\langle \bar{\theta} \rangle [S] = f_{\mathrm{S}}, \tag{6.101}$$

and the expression for the driving force (6.95) has the form

$$f_{\mathrm{S}} = -[\bar{W}] + \langle \bar{\sigma} \rangle [\bar{\varepsilon}] - \langle S \rangle [\bar{\theta}]. \tag{6.102}$$

Introducing the homothermal part of the driving force f_{H} as

$$f_{\mathrm{H}} = [-\bar{W} + \langle \bar{\sigma} \rangle \bar{\varepsilon}] = [\bar{\theta} S], \tag{6.103}$$

we can reduce the situation to that in the isothermal case. The difference between the adiabatic and isothermal cases is that the function f, constructed for the calculation of the entropy derivative, is connected here only with the homothermal part of the driving force

$$f_{\mathrm{H}} = [\bar{\theta} S] = [f], \quad \langle f \rangle = \langle \bar{\theta} S \rangle. \tag{6.104}$$

The expression for the function f in the martensitic part is modified accounting for the latent heat [Abeyaratne and Knowles (1993)]

$$
\begin{aligned}
f = -\frac{1}{2}\bar{\sigma}(\bar{\varepsilon} - \varepsilon_{tr}) &+ \frac{1}{2}\frac{C}{\theta_0}(\bar{\theta} - \theta_0)^2 - \frac{m}{2}(\bar{\theta} - \theta_0)(\bar{\varepsilon} - \varepsilon_{tr}) \\
&+ \sigma_{tr}\varepsilon_{tr} + \frac{L_{tr}}{\theta_0}(\bar{\theta} - \theta_0) + \langle \bar{\sigma} \rangle (\bar{\varepsilon} - \varepsilon_{tr}) + f_0,
\end{aligned}
\tag{6.105}
$$

where L_{tr} is the latent heat per unit volume. Correspondingly, in the austenitic phase

$$f = -\frac{1}{2}\bar{\sigma}\bar{\varepsilon} + \frac{1}{2}\frac{C}{\theta_0}(\bar{\theta} - \theta_0)^2 - \frac{m}{2}(\bar{\theta} - \theta_0)\bar{\varepsilon} + \langle \bar{\sigma} \rangle \bar{\varepsilon} + f_0. \tag{6.106}$$

The jump of the derivative of f with respect to the strain at fixed stress in both cases is given by

$$\left[\left(\frac{\partial f}{\partial \varepsilon}\right)_\sigma\right] = -\frac{1}{2}[\bar{\sigma}] - \frac{1}{2}\left[m(\bar{\theta} - \theta_0)\right], \tag{6.107}$$

because temperature is determined independently of other fields in the considered adiabatic approximation.

The combined jump relation determining the initiation of the phase transformation (6.51) then has the form

$$f_{\mathrm{H}} \left\langle \frac{(\lambda + 2\mu)}{\bar{\theta}m} \right\rangle + \langle f \rangle \left[\frac{(\lambda + 2\mu)}{\bar{\theta}m}\right] = -\frac{1}{2}[m(\bar{\theta} + \theta_0)]. \tag{6.108}$$

The expressions for the mean value and jump of the function f become dependent on the temperature

$$\langle f \rangle = \frac{1}{4}\left(\bar{\sigma}_M \bar{\varepsilon}_A + \bar{\sigma}_A(\bar{\varepsilon}_M - \varepsilon_{tr}) + 2\sigma_{tr}\varepsilon_{tr}\right) + f_0 + \langle f_\theta \rangle, \qquad (6.109)$$

$$[f] = \frac{1}{2}\left(\bar{\sigma}_M \bar{\varepsilon}_A - \bar{\sigma}_A(\bar{\varepsilon}_M - \varepsilon_{tr}) - 2\sigma_{tr}\varepsilon_{tr}\right) + [f_\theta], \qquad (6.110)$$

where the thermal part f_θ can be represented as

$$f_\theta = -\frac{m}{2}(\bar{\theta} - \theta_0)(\bar{\varepsilon} - \varepsilon_{tr}) + \frac{1}{2}\frac{C}{\theta_0}(\bar{\theta} - \theta_0)^2 + \frac{L_{tr}}{\theta_0}(\bar{\theta} - \theta_0), \qquad (6.111)$$

noting that the transformation strain and the latent heat are zero in the austenitic part.

We can rewrite the relations (6.109) and (6.110) as follows

$$\langle f \rangle - f_0 - \langle f_\theta \rangle - \frac{1}{2}\sigma_{tr}\varepsilon_{tr} = \frac{\bar{\sigma}_A \bar{\sigma}_M}{2}\left\langle \frac{1}{\lambda + 2\mu} \right\rangle, \qquad (6.112)$$

$$[f] - [f_\theta] + \sigma_{tr}\varepsilon_{tr} = \frac{\bar{\sigma}_A \bar{\sigma}_M}{2}\left[\frac{1}{\lambda + 2\mu}\right]. \qquad (6.113)$$

Dividing both parts of Eq. (6.112) by the corresponding parts of Eq. (6.113), we can eliminate stresses from the relation between mean value and jump of the function f

$$\frac{\langle f \rangle - f_0 - \langle f_\theta \rangle - \frac{1}{2}\sigma_{tr}\varepsilon_{tr}}{[f] - [f_\theta] + \sigma_{tr}\varepsilon_{tr}} = \left\langle \frac{1}{\lambda + 2\mu} \right\rangle \left[\frac{1}{\lambda + 2\mu}\right]^{-1}. \qquad (6.114)$$

After rearranging the last equation, we will have for the mean value of the function f

$$\langle f \rangle = f_0 + \langle f_\theta \rangle + \frac{1}{2}\sigma_{tr}\varepsilon_{tr}$$
$$+ ([f] - [f_\theta] + \sigma_{tr}\varepsilon_{tr})\left\langle \frac{1}{\lambda + 2\mu} \right\rangle \left[\frac{1}{\lambda + 2\mu}\right]^{-1}. \qquad (6.115)$$

Substituting the relation (6.115) into Eq. (6.108) and remembering that the jump of the function f is the driving force $f_{\mathbb{H}}$, we obtain

$$f_{\mathbb{H}}\left[\frac{1}{\bar{\theta}m}\right]\left[\frac{1}{\lambda + 2\mu}\right]^{-1} = -\frac{1}{2}[m(\bar{\theta} + \theta_0)]$$
$$- \left(f_0 + \frac{1}{2}\sigma_{tr}\varepsilon_{tr} + \sigma_{tr}\varepsilon_{tr}\left\langle \frac{1}{\lambda + 2\mu} \right\rangle \left[\frac{1}{\lambda + 2\mu}\right]^{-1}\right)\left[\frac{(\lambda + 2\mu)}{\bar{\theta}m}\right] \qquad (6.116)$$
$$- \left(\langle f_\theta \rangle \left[\frac{1}{\lambda + 2\mu}\right] - [f_\theta]\left\langle \frac{1}{\lambda + 2\mu} \right\rangle\right)\left[\frac{(\lambda + 2\mu)}{\bar{\theta}m}\right]\left[\frac{1}{(\lambda + 2\mu)}\right]^{-1}.$$

Therefore, the combined jump relation at the phase boundary (6.116) determines the value of the driving force at the interface. This value depends on the choice of the constant f_0. It is easy to see that the simplest expression can be obtained if we choose the value of f_0 as

$$f_0 = -\frac{1}{2}\sigma_{tr}\varepsilon_{tr} - \sigma_{tr}\varepsilon_{tr}\left\langle\frac{1}{\lambda+2\mu}\right\rangle\left[\frac{1}{\lambda+2\mu}\right]^{-1}. \qquad (6.117)$$

Since initial temperatures in both austenite and martensite are equal to the reference temperature θ_0, the thermal part of the function f is zero before phase transformation starts. This means that the initiation criterion for beginning of the phase transformation in the adiabatic case is identical to that in the isothermal case (6.54) as well as the choice of the constant f_0 (6.117)

$$|f_\mathbb{H}| \geq |f_{critical}|, \qquad (6.118)$$

where

$$f_{critical} = \theta_0^2[m]\left[\frac{1}{m}\right]^{-1}\left[\frac{1}{\lambda+2\mu}\right]. \qquad (6.119)$$

However, further motion of the phase boundary depends on the temperature. This leads to a more complicated expression for the critical value of the driving force in the adiabatic case. That is,

$$\begin{aligned}
f_{critical} &= \frac{1}{2}[m(\bar{\theta}+\theta_0)]\left[\frac{1}{\bar{\theta}m}\right]^{-1}\left[\frac{1}{\lambda+2\mu}\right] \\
&+ \left(\langle f_\theta\rangle\left[\frac{1}{\lambda+2\mu}\right] - [f_\theta]\left\langle\frac{1}{\lambda+2\mu}\right\rangle\right)\left[\frac{(\lambda+2\mu)}{\bar{\theta}m}\right]\left[\frac{1}{\bar{\theta}m}\right]^{-1}.
\end{aligned} \qquad (6.120)$$

Therefore, the critical value of the driving force can be calculated only numerically at each time step.

6.4.4 *Velocity of the phase boundary*

We are still able to express the stress jump at the moving phase boundary in terms of the driving force.

$$\begin{aligned}
[\bar{\sigma}] &= [m(\bar{\theta}+\theta_0)] - 2f_\mathbb{H}\left[\frac{1}{m}\right]\left[\frac{1}{\lambda+2\mu}\right]^{-1} \\
&-2\left(\langle f_\theta\rangle - [f_\theta]\left\langle\frac{1}{\lambda+2\mu}\right\rangle\left[\frac{1}{\lambda+2\mu}\right]^{-1}\right)\left[\frac{(\lambda+2\mu)}{\bar{\theta}m}\right].
\end{aligned} \qquad (6.121)$$

Having the value of the stress jump, we can determine the material velocity at the moving discontinuity by means of the jump relation for linear momentum (6.92)

$$\rho \bar{V}_N^2 [\bar{\varepsilon}] = [\bar{\sigma}]. \tag{6.122}$$

Relation (6.122) fully determines the velocity of the phase transition front propagation. However, we can express the influence of the temperature in a more explicit form. In the adiabatic case we have for the strain jump

$$[\bar{\varepsilon}] = \left[\frac{\bar{\sigma}}{\lambda + 2\mu} \right] - \left[\frac{m(\bar{\theta} - \theta_0)}{\lambda + 2\mu} \right] - \varepsilon_{tr}, \tag{6.123}$$

or

$$[\bar{\varepsilon}] = [\bar{\sigma}] \left\langle \frac{1}{\lambda + 2\mu} \right\rangle + \langle \bar{\sigma} \rangle \left[\frac{1}{\lambda + 2\mu} \right] - \left[\frac{m(\bar{\theta} - \theta_0)}{\lambda + 2\mu} \right] - \varepsilon_{tr}. \tag{6.124}$$

Representing the mean value of the stress as

$$\langle \bar{\sigma} \rangle = \bar{\sigma}_A - [\bar{\sigma}]/2, \tag{6.125}$$

we can rewrite the expression for the stress jump at the phase boundary as follows

$$[\bar{\varepsilon}] = A[\bar{\sigma}] - B(\bar{\theta}), \tag{6.126}$$

where

$$A = \left\langle \frac{1}{\lambda + 2\mu} \right\rangle - \frac{1}{2} \left[\frac{1}{\lambda + 2\mu} \right], \tag{6.127}$$

$$B(\bar{\theta}) = \varepsilon_{tr} - \bar{\sigma}_A \left[\frac{1}{\lambda + 2\mu} \right] + \left[\frac{m(\bar{\theta} - \theta_0)}{\lambda + 2\mu} \right]. \tag{6.128}$$

Therefore, Eq. (6.122) can be represented in terms of the stress jump

$$\rho_0 \bar{V}_N^2 = \frac{[\bar{\sigma}]}{A[\bar{\sigma}] - B(\bar{\theta})}. \tag{6.129}$$

As follows from Eq. (6.121), the stress jump is connected to the driving force by

$$[\bar{\sigma}] = D f_{\mathbb{H}} + H(\bar{\theta}), \tag{6.130}$$

where

$$D = -2 \left[\frac{1}{m} \right] \left[\frac{1}{\lambda + 2\mu} \right]^{-1}, \tag{6.131}$$

$$H(\bar{\theta}) = [m(\bar{\theta} + \theta_0)]$$

$$-2\left(\langle f_\theta \rangle - [f_\theta]\left\langle \frac{1}{\lambda + 2\mu} \right\rangle \left[\frac{1}{\lambda + 2\mu}\right]^{-1}\right)\left[\frac{(\lambda + 2\mu)}{\bar{\theta}m}\right]. \qquad (6.132)$$

This leads to the kinetic relation of the form

$$\rho_0 \bar{V}_N^2 = \frac{Df_{\mathbb{H}} + H(\bar{\theta})}{ADf_{\mathbb{H}} + AH(\bar{\theta}) - B(\bar{\theta})}, \qquad (6.133)$$

which should be modified taking into account the critical value of the driving force in the same way as in the isothermal case

$$\rho_0 \bar{V}_N^2 = \frac{D(f_{\mathbb{H}} - f_{critical}) + H(\bar{\theta})}{AD(f_{\mathbb{H}} - f_{critical}) + ADf_{critical} + AH(\bar{\theta}) - B(\bar{\theta})}. \qquad (6.134)$$

The temperature dependence allows us to calculate the velocity of the phase boundary in the adiabatic case only numerically.

6.4.5 Temperature field

The rate of entropy production due to the moving discontinuity is determined by [Abeyaratne and Knowles (2000)]

$$\frac{\partial S}{\partial t} = \frac{1}{V} \int_S \frac{f_S \bar{V}_N}{\langle \bar{\theta} \rangle} dA. \qquad (6.135)$$

For each individual computational cell n it follows that

$$\frac{\partial S_n}{\partial t} = \frac{f_S \bar{V}_N}{\langle \bar{\theta} \rangle \Delta x}. \qquad (6.136)$$

Therefore, the variation in entropy at each time step can be computed as

$$S_n^{k+1} - S_n^k = \frac{f_S \bar{V}_N \Delta t}{\langle \bar{\theta} \rangle \Delta x}. \qquad (6.137)$$

It follows that in the bulk, where the driving force is zero, we still have the adiabatic process

$$\bar{S}_n^{k+1} - \bar{S}_n^k = 0. \qquad (6.138)$$

If we take into account the latent heat, the entropy variation becomes

$$S_n^{k+1} - S_n^k = \frac{(f_S + L_{tr})\bar{V}_N \Delta t}{\langle \bar{\theta} \rangle \Delta x}. \qquad (6.139)$$

The corresponding heat supply reads

$$Q_n = \bar{\theta}_n(S_n^{k+1} - S_n^k) = \frac{\bar{\theta}_n f_{\mathbb{S}} \bar{V}_N \Delta t}{\langle \bar{\theta} \rangle \Delta x}. \tag{6.140}$$

From another side the same heat supply is connected with the temperature variation by

$$Q_n = C_n(\bar{\theta}_n^{k+1} - \bar{\theta}_n^k). \tag{6.141}$$

Consequently, we can calculate the temperature variation in terms of the driving force and the velocity of the discontinuity

$$\bar{\theta}_n^{k+1} - \bar{\theta}_n^k = \frac{\bar{\theta}_n(f_{\mathbb{S}} + L_{tr})\bar{V}_N \Delta t}{C_n \langle \bar{\theta} \rangle \Delta x}. \tag{6.142}$$

However, we have the possibility to determine only the value of the homothermal part of the driving force $f_{\mathbb{H}}$ whereas in the last expression we need to know the value of the total driving force $f_{\mathbb{S}}$. As we remember,

$$f_{\mathbb{H}} = [\bar{\theta}S], \quad f_{\mathbb{S}} = [S]\langle\bar{\theta}\rangle, \quad f = \bar{\theta}S. \tag{6.143}$$

Therefore, the correspondence between $f_{\mathbb{S}}$ and $f_{\mathbb{H}}$ is the following:

$$f_{\mathbb{S}} = \langle\bar{\theta}\rangle[S] = \langle\bar{\theta}\rangle\left[\frac{f}{\bar{\theta}}\right] = \langle\bar{\theta}\rangle f_{\mathbb{H}}\left\langle\frac{1}{\bar{\theta}}\right\rangle + \langle\bar{\theta}\rangle\langle f\rangle\left[\frac{1}{\bar{\theta}}\right]. \tag{6.144}$$

The final expression for the temperature field calculation becomes

$$\bar{\theta}_n^{k+1} - \bar{\theta}_n^k = \frac{\bar{\theta}_n \bar{V}_N \Delta t}{C_n \Delta x}\left(f_{\mathbb{H}}\left\langle\frac{1}{\bar{\theta}}\right\rangle + \langle f\rangle\left[\frac{1}{\bar{\theta}}\right] + \frac{L_{tr}}{\langle\bar{\theta}\rangle}\right). \tag{6.145}$$

It should be noted that temperature in the considered adiabatic case remains unchanged if the velocity of the discontinuity is zero.

6.5   Numerical simulations

The numerical procedure is the same as previously. This means that in addition to calculations by the wave-propagation algorithm in terms of excess quantities, a cellular automaton is used for the front motion due to the fixed grid. The phase state is moved to the next grid point only if the virtual displacement of the front is larger than the value of the space step. The temperature field is calculated as described in the previous section.

Fig. 6.13   One-dimensional front propagation in a bar.

6.5.1   *Pulse loading*

We consider a Ni-Ti bar subjected to a pulse loading. The geometry of the problem is shown in Fig. 6.13. The pulse loading is applied at the left boundary of the bar, the right boundary being fixed. The time history of the loading is presented in Fig. 6.14.

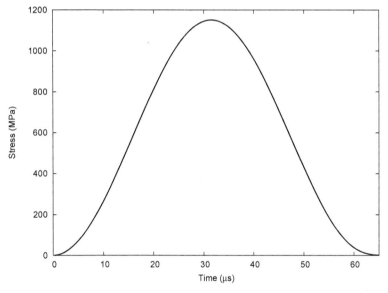

Fig. 6.14   Time history of pulse loading.

The amplitude of the loading was chosen to overcome the critical value of the driving force. Material properties for Ni-Ti SMA were extracted from the paper by McKelvey and Ritchie (2000). Simulations were performed for a bar of a length 18 cm at the temperature 22 °C. The Young's moduli are 62 GPa and 22 GPa for austenite and martensite phases, respectively, Poisson's ratio is the same for both phases and is equal to 0.33, the density

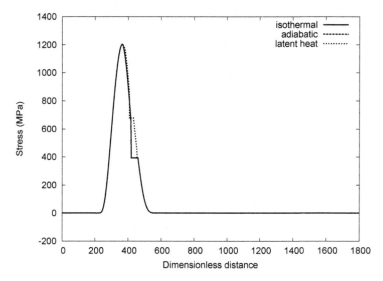

Fig. 6.15   Pulse shape at 212 µs.

for both phases is 6450 kg/m^3, the value of the transformation strain under uniaxial tension is 3.3 %. The initial location of the phase boundary was chosen at 400 space steps.

The goal of the simulation is to analyze the phase-transition front propagation under the dynamic loading including heating effects due to the entropy production at the moving phase boundary.

Snapshots of the pulse propagation at different time instants are shown in Figs. 6.15 – 6.18, where isothermal and adiabatic cases are presented simultaneously. As one can see, the overall behavior of the pulse in the adiabatic case is similar to that in the isothermal case [Berezovski and Maugin (2005b)]. However, the influence of latent heat leads to an increase of the transformation stress. If the effect of latent heat is not taken into account, there is almost no distinction in the pulse behavior in isothermal and in adiabatic cases.

6.5.2   Temperature distribution

The difference in the behavior of the pulse follows from the temperature variation, snapshots of which are shown in Figs. 6.19 – 6.22 for the corresponding time instants. The temperature variation in the adiabatic case without latent heat influence is only about 2 ℃, while the latent heat pro-

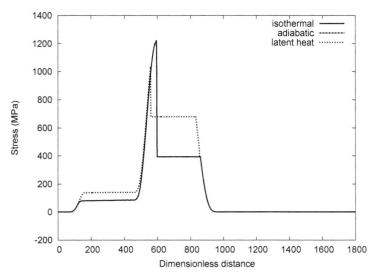

Fig. 6.16   Pulse shape at 318 μs.

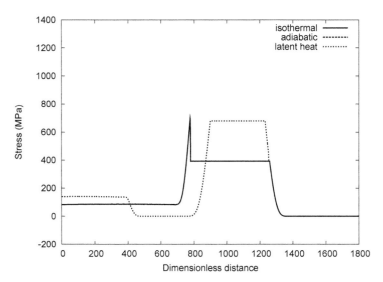

Fig. 6.17   Pulse shape at 424 μs.

vides an increase of the temperature for approximately 25 °C. The value of the latent heat for Ni-Ti was chosen as $L_{tr} = 8.5 \times 10^7$ J/m^3 [Messner and Werner (2003)], and the specific heat capacity $c_p = 500$ J/kgK [Vitiello,

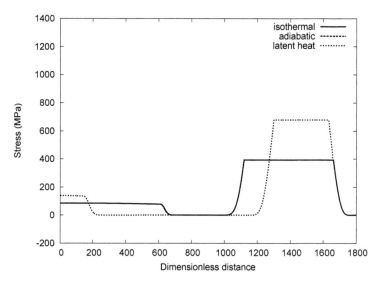

Fig. 6.18   Pulse shape at 530 μs.

Giorleo and Morace (2005)]. This means that the heating due to the entropy production at the moving phase boundary is much smaller than the heating due to the latent heat release in the considered Ni-Ti shape-memory alloy.

The location of the phase boundary corresponds to the right edge of the non-zero temperature in Figs. 6.19 - 6.22.

6.5.3   *Kinetic behavior*

The temperature variation affects also the kinetic behavior of the phase boundary. This is illustrated in Fig. 6.23, where the velocity of the phase boundary versus the driving force is presented.

Values of the velocity of the phase boundary are calculated at each time step by means of Eq. (6.122) and then normalized with respect to the value of sound velocity in austenite. The corresponding driving force $f_{\mathbb{H}}$ (6.103) is computed using relations for the free energy density (6.86) and (6.87) for martensitic and austenitic phases, respectively.

It should be noted that the initial kinetic behavior both with and without latent heat influence is the same, because there is no initial temperature variation. After the beginning of heating due to the entropy production at the moving phase boundary, the influence of latent heat results in the

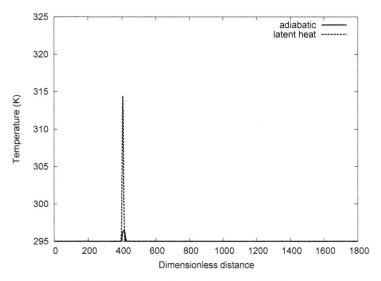

Fig. 6.19   Temperature distributions at 212 μs.

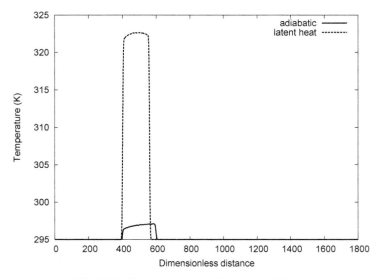

Fig. 6.20   Temperature distributions at 318 μs.

change of the kinetic behavior of the phase boundary from its adiabatic curve without latent heat influence. Further the kinetic curves remain different.

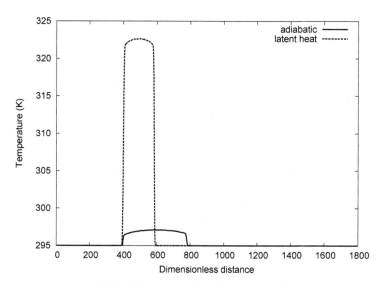

Fig. 6.21   Temperature distributions at 424 μs.

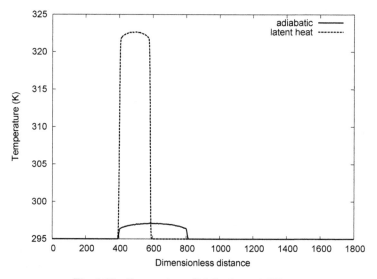

Fig. 6.22   Temperature distributions at 530 μs.

Small values of the velocity of the phase boundary at the kinetic curve with the latent heat influence correspond to late stages of the process, when the driving force decreases following the diminishing of the pulse

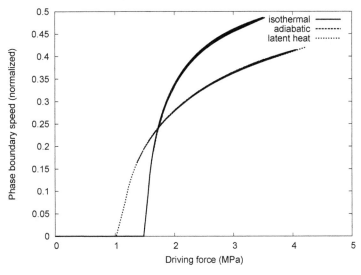

Fig. 6.23   Kinetic relations.

loading. This means that zero value of the velocity of the phase boundary at the curve with the latent heat influence relates to the end of the phase transformation process, but not to its beginning.

6.6   Concluding remarks

The influence of temperature variation on the propagation of martensitic phase-transition front is studied on the example of one-dimensional impact problem for a Ni-Ti bar. The adiabatic description is chosen since the heat conduction is insignificant because of rapid loading. The presented theory generalizes the isothermal description presented before [Berezovski and Maugin (2005b)].

In the considered adiabatic formulation, the excess stresses needed for numerical calculations are determined in the same way as in the isothermal case. The local equilibrium jump relations provide the possibility to establish a kinetic relation taking into account the temperature variation. The corresponding initiation criterion of phase transformation is also more general than that in the isothermal case.

Both the kinetic relation and the initiation criterion are derived under assumption of the continuity of excess stresses at the phase boundary. This assumption can be a subject of generalization. The choice of the constant f_0, which leads to the elimination of the influence of the transformation strain on the critical value of the driving force, can also be changed to follow experimental data.

Numerical simulations of the phase-transition front propagation in Ni-Ti bar show that the entropy production at the moving phase boundary and latent heat release provide two different contributions in the temperature variation. The latter leads to distinct kinetic behaviors of the phase boundary depending on the latent heat release.

Chapter 7

Two-Dimensional Elastic Waves in Inhomogeneous Media

The use of simplified one-dimensional models should always be accompanied by questions about the accuracy of geometrical assumptions. On the other hand, the more complicated geometry of two- and three-dimensional models clearly needs more sophisticated numerical algorithms and surely much larger computational capacities. In this chapter we start analysis of two-dimensional problems from presenting a stable algorithm for numerical calculations and demonstrate its effectiveness with several examples.

A common approach when solving multi-dimensional hyperbolic problems is to apply dimensional splitting [LeVeque (2002a)]. The idea is to iterate on one-dimensional problems. However, it is well known that dimensional splitting has several disadvantages. Since the strategy only involves flow in the coordinate directions, the solution is affected by the grid orientation. Discontinuities travelling obliquely to the grid experience more smearing than those travelling in the coordinate directions.

In unsplit methods, information is propagated in a multi-dimensional way. The modification of the unsplit scheme [LeVeque (1997)] is applied below. One-dimensional Riemann problems are solved at the interfaces. The left-going and right-going waves are split into parts propagating in the transverse direction by solving Riemann problems in coordinate directions tangential to the interfaces. This models cross-derivative terms necessary for obtaining both a stable and formally second order scheme. The same numerical scheme is applied for wave and front propagation.

7.1   Governing equations

The governing equations of inhomogeneous linear isotropic elasticity can be represented by

$$\rho_0(\mathbf{x})\frac{\partial v_i}{\partial t} = \frac{\partial \sigma_{ij}}{\partial x_j}, \qquad (7.1)$$

$$\frac{\partial \sigma_{ij}}{\partial t} = \lambda(\mathbf{x})\frac{\partial v_k}{\partial x_k}\delta_{ij} + \mu(\mathbf{x})\left(\frac{\partial v_i}{\partial x_j} + \frac{\partial v_j}{\partial x_i}\right). \qquad (7.2)$$

If we divide the body in a finite number of identical elements of elementary volume ΔV and boundary $\partial \Delta V$, the integration over the finite volume element of Eqs. (7.1) - (7.2) and of the definition of the strain rate yields the following set of integral forms:

$$\frac{\partial}{\partial t}\int_{\Delta V}\rho_0 v_i dV = \int_{\partial \Delta V}\sigma_{ij}n_j dA, \qquad (7.3)$$

$$\frac{\partial}{\partial t}\int_{\Delta V}\varepsilon_{ij} dV = \int_{\partial \Delta V}h_{ijk}n_k dA, \qquad (7.4)$$

$$\frac{\partial}{\partial t}\int_{\Delta V}\sigma_{ij} dV = \int_{\partial \Delta V}(2\mu h_{ijk}n_k + \lambda\delta_{ij}v_k n_k)dA + \varphi_{ij}, \qquad (7.5)$$

where $h_{ijk} = 1/2(\delta_{ik}v_j + \delta_{jk}v_i)$, n_i is the unit outward normal to the boundary of a discrete element, and source terms due to material inhomogeneities are given by

$$\varphi_{ij}^{inh} = -\int_{\Delta V}\left(v_k\frac{\partial \lambda}{\partial x_k}\delta_{ij} + v_i\frac{\partial \mu}{\partial x_j} + v_j\frac{\partial \mu}{\partial x_i}\right)dV. \qquad (7.6)$$

7.1.1   *Averaged quantities*

Introducing averaged quantities at each time step

$$\bar{v}_i = \frac{1}{\Delta V}\int_{\Delta V}v_i dV, \qquad \bar{\varepsilon}_{ij} = \frac{1}{\Delta V}\int_{\Delta V}\varepsilon_{ij} dV, \qquad (7.7)$$

$$\bar{\sigma}_{ij} = \frac{1}{\Delta V}\int_{\Delta V}\sigma_{ij} dV, \qquad (7.8)$$

and numerical fluxes at the boundaries of each element

$$F_{ij} \approx \frac{1}{\Delta t} \int_{t_l}^{t_{l+1}} \sigma_{ij}|_{\partial \Delta V} \, dt, \tag{7.9}$$

$$G_{ijk} \approx \frac{1}{\Delta t} \int_{t_l}^{t_{l+1}} (2\mu h_{ijk} + \lambda \delta_{ij} v_k)|_{\partial \Delta V} \, dt, \tag{7.10}$$

we are able to write a finite volume numerical scheme for Eqs. (7.3), (7.5) for a uniform grid of homogeneous elements in the form

$$(\bar{v}_i)^{l+1} - (\bar{v}_i)^l = \frac{\Delta t}{\rho \Delta x_j} \left((F_{ij}^{in})^l + (F_{ij}^{out})^l \right), \tag{7.11}$$

and

$$(\bar{\sigma}_{ij})^{l+1} - (\bar{\sigma}_{ij})^l = \frac{\Delta t}{\Delta x_k} \left((G_{ijk}^{in})^l + (G_{ijk}^{out})^l \right), \tag{7.12}$$

where l denotes time steps and superscripts "in" and "out" denote inflow and outflow parts in the flux decomposition.

The main difficulty in the construction of a numerical scheme is the proper determination of the numerical fluxes [LeVeque (2002a)]. Since numerical simulations are performed for the two-dimensional case, we consider this case in more detail.

7.1.2 Conservation law

In two space dimensions, the system of equations for linear elasticity in an inhomogeneous medium, i.e.,

$$\rho(x,y)\frac{\partial v_1}{\partial t} = \frac{\partial \sigma_{11}}{\partial x} + \frac{\partial \sigma_{12}}{\partial y}, \tag{7.13}$$

$$\rho(x,y)\frac{\partial v_2}{\partial t} = \frac{\partial \sigma_{21}}{\partial x} + \frac{\partial \sigma_{22}}{\partial y}, \tag{7.14}$$

$$\frac{\partial \sigma_{11}}{\partial t} = (\lambda(x,y) + 2\mu)(x,y)\frac{\partial v_1}{\partial x} + \lambda(x,y)\frac{\partial v_2}{\partial y}, \tag{7.15}$$

$$\frac{\partial \sigma_{22}}{\partial t} = \lambda(x,y)\frac{\partial v_1}{\partial x} + (\lambda(x,y) + 2\mu(x,y))\frac{\partial v_2}{\partial y}, \tag{7.16}$$

$$\frac{\partial \sigma_{12}}{\partial t} = \frac{\partial \sigma_{21}}{\partial t} = \mu(x,y)\left(\frac{\partial v_1}{\partial y} + \frac{\partial v_2}{\partial x} \right), \tag{7.17}$$

can also be represented in the matrix form

$$\frac{\partial q}{\partial t} = A(x,y)\,\frac{\partial q}{\partial x} + B(x,y)\,\frac{\partial q}{\partial y}, \qquad (7.18)$$

where

$$q = \begin{bmatrix} \sigma_{11} \\ \sigma_{12} \\ \sigma_{22} \\ v_1 \\ v_2 \end{bmatrix}, \qquad (7.19)$$

$$A = \begin{bmatrix} 0 & 0 & 0 & -\lambda(x,y) - 2\,\mu(x,y) & 0 \\ 0 & 0 & 0 & 0 & -\mu(x,y) \\ 0 & 0 & 0 & -\lambda(x,y) & 0 \\ -\dfrac{1}{\rho(x,y)} & 0 & 0 & 0 & 0 \\ 0 & -\dfrac{1}{\rho(x,y)} & 0 & 0 & 0 \end{bmatrix}, \qquad (7.20)$$

$$B = \begin{bmatrix} 0 & 0 & 0 & 0 & -\lambda(x,y) \\ 0 & 0 & 0 & -\mu(x,y) & 0 \\ 0 & 0 & 0 & 0 & -\lambda(x,y) - 2\,\mu(x,y) \\ 0 & -\dfrac{1}{\rho(x,y)} & 0 & 0 & 0 \\ 0 & 0 & -\dfrac{1}{\rho(x,y)} & 0 & 0 \end{bmatrix}. \qquad (7.21)$$

The wave-propagation algorithm for the solution of these equations is based on solving Riemann problems at the interface between grid cells to determine the flux decomposition [LeVeque (1997)]. The fluctuations arising from Riemann problems in the x- and y-directions, respectively, are determined exactly by means of eigenvalues and eigenvectors of matrices A and B [LeVeque (1997)].

7.2   Fluctuation splitting

The space coordinates are discretized with uniform spacing Δx and Δy. We assume that the cell (ij) has constant (or averaged) material parameters

ρ_{ij}, λ_{ij}, and μ_{ij} and set local matrices A_{ij} and B_{ij}, where coefficients depend on the state of the cell (ij).

For all ij, the eigenvalues of the matrix A_{ij} are the following

$$\alpha_{ij}^{(1)} = -c_{Pij}, \; \alpha_{ij}^{(2)} = c_{Pij}, \; \alpha_{ij}^{(3)} = -c_{Sij}, \; \alpha_{ij}^{(4)} = c_{Sij}, \; \alpha_{ij}^{(5)} = 0, \quad (7.22)$$

where $c_{Sij} = \sqrt{\mu_{ij}/\rho_{ij}}$ and $c_{Pij} = \sqrt{(\lambda_{ij} + 2\mu_{ij})/\rho_{ij}}$.

Corresponding eigenvectors can be written as follows

$$r_{ij}^{(1)} = \begin{bmatrix} c_{Pij} \\ 0 \\ \lambda_{ij}/\rho_{ij}c_{Pij} \\ 1/\rho_{ij} \\ 0 \end{bmatrix}, \qquad r_{ij}^{(2)} = \begin{bmatrix} -c_{Pij} \\ 0 \\ -\lambda_{ij}/\rho_{ij}c_{Pij} \\ 1/\rho_{ij} \\ 0 \end{bmatrix}, \qquad (7.23)$$

$$r_{ij}^{(3)} = \begin{bmatrix} 0 \\ c_{Sij} \\ 0 \\ 0 \\ 1/\rho_{ij} \end{bmatrix}, \qquad r_{ij}^{(4)} = \begin{bmatrix} 0 \\ -c_{Sij} \\ 0 \\ 0 \\ 1/\rho_{ij} \end{bmatrix}, \qquad r_{ij}^{(5)} = \begin{bmatrix} 0 \\ 0 \\ 1 \\ 0 \\ 0 \end{bmatrix}. \qquad (7.24)$$

The fluctuations between states $q_{i-1\,j}$ and q_{ij} consist of five waves, but one of them always has a zero speed and can be ignored [LeVeque (1997)]

$$\mathcal{A}^- \Delta q_{ij} = -c_{Pi-1\,j}\gamma_{ij}^{(1)} r_{i-1\,j}^{(1)} - c_{Si-1\,j}\gamma_{ij}^{(3)} r_{i-1\,j}^{(3)}, \qquad (7.25)$$

$$\mathcal{A}^+ \Delta q_{ij} = c_{Pij}\gamma_{ij}^{(2)} r_{ij}^{(2)} + c_{Sij}\gamma_{ij}^{(4)} r_{ij}^{(4)}. \qquad (7.26)$$

The coefficients $\gamma_{ij}^{(*)}$ are determined by the exact solution of the corresponding Riemann problem

$$\left(\frac{c_{Pi-1\,j}}{\rho_{ij}} + \frac{c_{Pij}}{\rho_{i-1\,j}} \right) \gamma_{ij}^{(1)}$$

$$(7.27)$$

$$= \left(\frac{(\bar{\sigma}_{11})_{ij} - (\bar{\sigma}_{11})_{i-1\,j}}{\rho_{ij}} + c_{Pij}\left[(\bar{v}_1)_{ij} - (\bar{v}_1)_{i-1\,j} \right] \right),$$

$$\left(\frac{c_{Pi-1\,j}}{\rho_{ij}} + \frac{c_{Pij}}{\rho_{i-1\,j}} \right) \gamma_{ij}^{(2)}$$

$$(7.28)$$

$$= \left(-\frac{(\bar{\sigma}_{11})_{ij} - (\bar{\sigma}_{11})_{i-1\,j}}{\rho_{i-1\,j}} + c_{Pi-1\,j}\left[(\bar{v}_1)_{ij} - (\bar{v}_1)_{i-1\,j} \right] \right),$$

$$\left(\frac{c_{Si-1\,j}}{\rho_{ij}} + \frac{c_{Sij}}{\rho_{i-1\,j}} \right) \gamma_{ij}^{(3)}$$

$$= \left(\frac{(\bar{\sigma}_{12})_{ij} - (\bar{\sigma}_{12})_{i-1\,j}}{\rho_{ij}} + c_{Sij} \left[(\bar{v}_2)_{ij} - (\bar{v}_2)_{i-1\,j} \right] \right), \tag{7.29}$$

$$\left(\frac{c_{Si-1\,j}}{\rho_{ij}} + \frac{c_{Sij}}{\rho_{i-1\,j}} \right) \gamma_{ij}^{(4)}$$

$$= \left(-\frac{(\bar{\sigma}_{12})_{ij} - (\bar{\sigma}_{12})_{i-1\,j}}{\rho_{i-1\,j}} + c_{Si-1\,j} \left[(\bar{v}_2)_{ij} - (\bar{v}_2)_{i-1\,j} \right] \right). \tag{7.30}$$

Accordingly, the eigenvalues of the matrix B_{ij} have the values

$$\beta_{ij}^{(1)} = -c_{Pij}, \ \beta_{ij}^{(2)} = c_{Pij}, \ \beta_{ij}^{(3)} = -c_{Sij}, \ \beta_{ij}^{(4)} = c_{Sij}, \ \beta_{ij}^{(5)} = 0, \tag{7.31}$$

with corresponding eigenvectors

$$s_{ij}^{(1)} = \begin{bmatrix} \lambda_{ij}/\rho_{ij}c_{Pij} \\ 0 \\ c_{Pij} \\ 0 \\ 1/\rho_{ij} \end{bmatrix}, \qquad s_{ij}^{(2)} = \begin{bmatrix} -\lambda_{ij}/\rho_{ij}c_{Pij} \\ 0 \\ -c_{Pij} \\ 0 \\ 1/\rho_{ij} \end{bmatrix}, \tag{7.32}$$

$$s_{ij}^{(3)} = \begin{bmatrix} 0 \\ c_{Sij} \\ 0 \\ 1/\rho_{ij} \\ 0 \end{bmatrix}, \qquad s_{ij}^{(4)} = \begin{bmatrix} 0 \\ -c_{Sij} \\ 0 \\ 1/\rho_{ij} \\ 0 \end{bmatrix}, \qquad s_{ij}^{(5)} = \begin{bmatrix} 1 \\ 0 \\ 0 \\ 0 \\ 0 \end{bmatrix}. \tag{7.33}$$

The fluctuations between states $q_{i\,j-1}$ and q_{ij} are determined similarly

$$\mathcal{B}^- \Delta q_{ij} = -c_{Pi\,j-1}\delta_{ij}^{(1)} s_{i\,j-1}^{(1)} - c_{Si,j-1}\delta_{ij}^{(3)} s_{i\,j-1}^{(3)}, \tag{7.34}$$

$$\mathcal{B}^+ \Delta q_{ij} = c_{Pij}\delta_{ij}^{(2)} s_{ij}^{(2)} + c_{Sij}\delta_{ij}^{(4)} s_{ij}^{(4)}, \tag{7.35}$$

where coefficients $\delta_{ij}^{(*)}$ are determined as follows

$$\left(\frac{c_{P\,i\,j-1}}{\rho_{ij}} + \frac{c_{Pij}}{\rho_{i\,j-1}} \right) \delta_{ij}^{(1)}$$

$$\tag{7.36}$$

$$= \left(\frac{(\bar{\sigma}_{22})_{i\,j} - (\bar{\sigma}_{22})_{i\,j-1}}{\rho_{ij}} + c_{Pij} \left[(\bar{v}_2)_{ij} - (\bar{v}_2)_{i\,j-1} \right] \right),$$

$$\left(\frac{c_{P\,i\,j-1}}{\rho_{ij}} + \frac{c_{Pij}}{\rho_{i\,j-1}} \right) \delta_{ij}^{(2)}$$

$$\tag{7.37}$$

$$= \left(-\frac{(\bar{\sigma}_{22})_{ij} - (\bar{\sigma}_{22})_{i\,j-1}}{\rho_{i\,j-1}} + c_{P\,i\,j-1} \left[(\bar{v}_2)_{ij} - (\bar{v}_2)_{i\,j-1} \right] \right),$$

$$\left(\frac{c_{S\,i\,j-1}}{\rho_{ij}} + \frac{c_{Sij}}{\rho_{i\,j-1}} \right) \delta_{ij}^{(3)}$$

$$\tag{7.38}$$

$$= \left(\frac{(\bar{\sigma}_{12})_{ij} - (\bar{\sigma}_{12})_{i\,j-1}}{\rho_{ij}} + c_{Sij} \left[(\bar{v}_1)_{ij} - (\bar{v}_1)_{i\,j-1} \right] \right),$$

$$\left(\frac{c_{S\,i\,j-1}}{\rho_{ij}} + \frac{c_{Sij}}{\rho_{i\,j-1}} \right) \delta_{ij}^{(4)}$$

$$\tag{7.39}$$

$$= \left(-\frac{(\bar{\sigma}_{12})_{ij} - (\bar{\sigma}_{12})_{i\,j-1}}{\rho_{i\,j-1}} + c_{S\,i\,j-1} \left[(\bar{v}_1)_{ij} - (\bar{v}_1)_{i\,j-1} \right] \right).$$

The construction of the wave-propagation algorithm begins with establishing the first-order Godunov scheme in terms of these fluctuations.

7.3 First-order Godunov scheme

The extension of the first-order Godunov scheme to two space dimensions has the following form [LeVeque (1997)]

$$\bar{q}_{ij}^{k+1} - \bar{q}_{ij}^{k} = -\frac{\Delta t}{\Delta x} (\mathcal{A}^+ \Delta q_{ij} + \mathcal{A}^- \Delta q_{i+1\,j}) - \frac{\Delta t}{\Delta y} (\mathcal{B}^+ \Delta q_{ij} + \mathcal{B}^- \Delta q_{i\,j+1}).$$

$$\tag{7.40}$$

The fluctuations arising from Riemann problems in the x- and y-directions, respectively, can be represented in terms of waves

$$\mathcal{A}^- \Delta q_{ij} = \sum_{p=1} \alpha_{ij}^{(p-)} W^{(p-)} \qquad \mathcal{A}^+ \Delta q_{ij} = \sum_{p=1} \alpha_{ij}^{(p+)} W^{(p+)}, \tag{7.41}$$

$$\mathcal{B}^- \Delta q_{ij} = \sum_{p=1} \beta_{ij}^{(p-)} w^{(p-)}, \qquad \mathcal{B}^+ \Delta q_{ij} = \sum_{p=1} \beta_{ij}^{(p+)} w^{(p+)}. \qquad (7.42)$$

Here $W^{(p)}$ and $w^{(p)}$ are horizontal and vertical waves corresponding to the local Riemann problem, superscripts "+" and "−" denoting positive and negative eigenvalues.

As in the one-dimensional case (Eqs. (4.56) and (4.57)), this algorithm can be rewritten in terms of excess quantities [Berezovski and Maugin (2001)]

$$(\bar{v}_1)_{ij}^{k+1} - (\bar{v}_1)_{ij}^{k}$$
$$= \frac{\Delta t}{\Delta x} \frac{1}{\rho_{ij}} \left[(\Sigma_{11}^+)_{ij}^k - (\Sigma_{11}^-)_{ij}^k \right] + \frac{\Delta t}{\Delta y} \frac{1}{\rho_{ij}} \left[(\Sigma_{12}^+)_{ij}^k - (\Sigma_{12}^-)_{ij}^k \right], \qquad (7.43)$$

$$(\bar{v}_2)_{ij}^{k+1} - (\bar{v}_2)_{ij}^{k}$$
$$= \frac{\Delta t}{\Delta x} \frac{1}{\rho_{ij}} \left[(\Sigma_{21}^+)_{ij}^k - (\Sigma_{21}^-)_{ij}^k \right] + \frac{\Delta t}{\Delta y} \frac{1}{\rho_{ij}} \left[(\Sigma_{22}^+)_{ij}^k - (\Sigma_{22}^-)_{ij}^k \right], \qquad (7.44)$$

$$(\bar{\sigma}_{11})_{ij}^{k+1} - (\bar{\sigma}_{11})_{ij}^{k}$$
$$= \frac{\Delta t}{\Delta x} (\lambda_{ij} + 2\mu_{ij}) \left[(V_{11}^+)_{ij}^k - (V_{11}^-)_{ij}^k \right] + \frac{\Delta t}{\Delta y} \lambda_{ij} \left[(V_{22}^+)_{ij}^k - (V_{22}^-)_{ij}^k \right], \qquad (7.45)$$

$$(\bar{\sigma}_{12})_{ij}^{k+1} - (\bar{\sigma}_{12})_{ij}^{k}$$
$$= \frac{\Delta t}{\Delta x} \mu_{ij} \left[(V_{21}^+)_{ij}^k - (V_{21}^-)_{ij}^k \right] + \frac{\Delta t}{\Delta y} \mu_{ij} \left[(V_{12}^+)_{ij}^k - (V_{12}^-)_{ij}^k \right], \qquad (7.46)$$

$$(\bar{\sigma}_{22})_{ij}^{k+1} - (\bar{\sigma}_{22})_{ij}^{k}$$
$$= \frac{\Delta t}{\Delta x} \lambda_{ij} \left[(V_{11}^+)_{ij}^k - (V_{11}^-)_{ij}^k \right] + \frac{\Delta t}{\Delta y} (\lambda_{ij} + 2\mu_{ij}) \left[(V_{22}^+)_{ij}^k - (V_{22}^-)_{ij}^k \right], \qquad (7.47)$$

where the values of excess quantities are calculated as follows

$$(V_{11}^+)_{ij}^k = \frac{\gamma_{i+1\,j}^{(1)}}{\rho_{ij}}, \qquad (V_{11}^-)_{ij}^k = -\frac{\gamma_{ij}^{(2)}}{\rho_{ij}}, \qquad (7.48)$$

$$(V_{22}^+)_{ij}^k = \frac{\delta_{i\,j+1}^{(1)}}{\rho_{ij}}, \qquad (V_{22}^-)_{ij}^k = -\frac{\delta_{ij}^{(2)}}{\rho_{ij}}, \qquad (7.49)$$

$$(V_{21}^+)_{ij}^k = \frac{\gamma_{i+1\,j}^{(3)}}{\rho_{ij}}, \qquad (V_{21}^-)_{ij}^k = -\frac{\gamma_{ij}^{(4)}}{\rho_{ij}}, \qquad (7.50)$$

$$(V_{12}^+)_{ij}^k = \frac{\delta_{i\,j+1}^{(3)}}{\rho_{ij}}, \qquad (V_{12}^-)_{ij}^k = -\frac{\delta_{ij}^{(4)}}{\rho_{ij}}, \qquad (7.51)$$

$$(\Sigma_{11}^+)_{ij}^k = c_{Pij}\gamma_{i+1\,j}^{(1)}, \qquad (\Sigma_{11}^-)_{ij}^k = c_{Pij}\gamma_{ij}^{(2)}, \qquad (7.52)$$

$$(\Sigma_{21}^+)_{ij}^k = c_{Sij}\gamma_{i+1\,j}^{(3)}, \qquad (\Sigma_{21}^-)_{ij}^k = c_{Sij}\gamma_{ij}^{(4)}, \qquad (7.53)$$

$$(\Sigma_{12}^+)_{ij}^k = c_{Sij}\delta_{i\,j+1}^{(3)}, \qquad (\Sigma_{12}^-)_{ij}^k = c_{Sij}\delta_{ij}^{(4)}, \qquad (7.54)$$

$$(\Sigma_{22}^+)_{ij}^k = c_{Pij}\delta_{i\,j+1}^{(1)}, \qquad (\Sigma_{22}^-)_{ij}^k = c_{Pij}\delta_{ij}^{(2)}. \qquad (7.55)$$

Godunov's method (7.40) is extended to a high resolution method by adding correction terms [LeVeque (1997)]. The form of the extended method is

$$\bar{q}_{ij}^{k+1} = \bar{q}_{ij}^k + \Delta_{ij}^{up} - \frac{\Delta t}{\Delta x}(\bar{F}_{i+1\,j}^k - \bar{F}_{ij}^k) - \frac{\Delta t}{\Delta y}(\bar{G}_{i\,j+1}^k - \bar{G}_{ij}^k), \qquad (7.56)$$

where Δ_{ij}^{up} is the update for a first-order upwind Godunov method (right-hand side of Eq. (7.40)). The second-order correction terms can be represented in terms of waves [LeVeque (1997)] by

$$\bar{F}_{ij}^k = \frac{1}{2}\sum_{p=1}^{4}|\alpha_{ij}^{(p)}|\left(1 - \frac{\Delta t}{\Delta x}|\alpha_{ij}^{(p)}|\right)W^{(p)}, \qquad (7.57)$$

$$\bar{G}_{ij}^k = \frac{1}{2}\sum_{p=1}^{4}|\beta_{ij}^{(p)}|\left(1 - \frac{\Delta t}{\Delta x}|\beta_{ij}^{(p)}|\right)w^{(p)}, \qquad (7.58)$$

and provide second-order accuracy. The obtained algorithm is a variant of the well-known Lax-Wendroff method [Lax and Wendroff (1964)].

The above described Godunov method is based on propagating waves normal to each cell interface. In reality the waves should propagate in a multi-dimensional manner and affect other cell averages besides those adjacent to the interface.

7.4   Transverse propagation

7.4.1   Vertical transverse propagation

The multidimensional motion is accomplished by splitting each fluctuation $\mathcal{A}^*\Delta q_{ij}$ into two transverse fluctuations, which will be called $\mathcal{B}^+\mathcal{A}^*\Delta q_{ij}$

(the up-going transverse fluctuation) and $\mathcal{B}^-\mathcal{A}^*\Delta q_{ij}$ (the down-going transverse fluctuation) [LeVeque (1997)] and such that

$$\mathcal{B}^-\mathcal{A}^-\Delta q_{ij} = -c_{Pij-1}(\beta^{(1)})^-_{ij}\, s^{(1)}_{ij-1} - c_{Sij-1}(\beta^{(3)})^-_{ij}\, s^{(3)}_{ij-1}, \qquad (7.59)$$

$$\mathcal{B}^+\mathcal{A}^-\Delta q_{ij} = c_{Pij}(\beta^{(2)})^-_{ij}\, s^{(2)}_{ij} + c_{Sij}(\beta^{(4)})^-_{ij}\, s^{(4)}_{ij}, \qquad (7.60)$$

$$\mathcal{B}^-\mathcal{A}^+\Delta q_{ij} = -c_{Pij-1}(\beta^{(1)})^+_{ij}\, s^{(1)}_{ij-1} - c_{Sij-1}(\beta^{(3)})^+_{ij}\, s^{(3)}_{ij-1}, \qquad (7.61)$$

$$\mathcal{B}^+\mathcal{A}^+\Delta q_{ij} = c_{Pij}(\beta^{(2)})^+_{ij}\, s^{(2)}_{ij} + c_{Sij}(\beta^{(4)})^+_{ij}\, s^{(4)}_{ij}, \qquad (7.62)$$

where the coefficients are calculated algebraically by means of the same values of coefficients $\gamma^{(*)}_{ij}$, which are obtained in the solution of the corresponding Riemann problems:

$$\begin{aligned}
&\left(\frac{c_{Pij-1}}{\rho_{ij}} + \frac{c_{Pij}}{\rho_{ij-1}}\right)(\beta^{(1)})^-_{ij} \\
&= \left(-\frac{c_{Si-1j}c_{Pij}}{\rho_{i-1j}}\gamma^{(3)}_{ij} - \frac{\lambda_{i-1j}}{\rho_{i-1j}\rho_{ij}}\gamma^{(1)}_{ij}\right),
\end{aligned} \qquad (7.63)$$

$$\begin{aligned}
&\left(\frac{c_{Pij-1}}{\rho_{ij}} + \frac{c_{Pij}}{\rho_{ij-1}}\right)(\beta^{(2)})^-_{ij} \\
&= \left(-\frac{c_{Si-1j}c_{Pij-1}}{\rho_{i-1j}}\gamma^{(3)}_{ij} + \frac{\lambda_{i-1j}}{\rho_{i-1j}\rho_{ij-1}}\gamma^{(1)}_{ij}\right),
\end{aligned} \qquad (7.64)$$

$$\begin{aligned}
&\left(\frac{c_{Sij-1}}{\rho_{ij}} + \frac{c_{Sij}}{\rho_{ij-1}}\right)(\beta^{(3)})^-_{ij} \\
&= \left(-\frac{c_{Si-1j}c_{Si-1j}}{\rho_{ij}}\gamma^{(3)}_{ij} - \frac{c_{Pi-1j}c_{Sij}}{\rho_{i-1j}}\gamma^{(1)}_{ij}\right),
\end{aligned} \qquad (7.65)$$

$$\begin{aligned}
&\left(\frac{c_{Sij-1}}{\rho_{ij}} + \frac{c_{Sij}}{\rho_{ij-1}}\right)(\beta^{(4)})^-_{ij} \\
&= \left(\frac{c_{Si-1j}c_{Si-1j}}{\rho_{ij-1}}\gamma^{(3)}_{ij} - \frac{c_{Pi-1j}c_{Sij-1}}{\rho_{i-1j}}\gamma^{(1)}_{ij}\right),
\end{aligned} \qquad (7.66)$$

$$\begin{aligned}
&\left(\frac{c_{Pij-1}}{\rho_{ij}} + \frac{c_{Pij}}{\rho_{ij-1}}\right)(\beta^{(1)})^+_{ij} \\
&= \left(\frac{c_{Sij}c_{Pij}}{\rho_{ij}}\gamma^{(4)}_{ij} - \frac{\lambda_{ij}}{\rho_{ij}\rho_{ij}}\gamma^{(2)}_{ij}\right),
\end{aligned} \qquad (7.67)$$

$$\left(\frac{c_{Pij-1}}{\rho_{ij}} + \frac{c_{Pij}}{\rho_{ij-1}} \right) (\beta^{(2)})_{ij}^{+}$$
$$= \left(\frac{c_{Sij}c_{Pij-1}}{\rho_{ij}} \gamma_{ij}^{(4)} + \frac{\lambda_{ij}}{\rho_{ij}\rho_{ij-1}} \gamma_{ij}^{(2)} \right),$$
(7.68)

$$\left(\frac{c_{Sij-1}}{\rho_{ij}} + \frac{c_{Sij}}{\rho_{ij-1}} \right) (\beta^{(3)})_{ij}^{+}$$
$$= \left(-\frac{c_{Sij}c_{Sij}}{\rho_{ij}} \gamma_{ij}^{(4)} + \frac{c_{Pij}c_{Sij}}{\rho_{ij}} \gamma_{ij}^{(2)} \right),$$
(7.69)

$$\left(\frac{c_{Sij-1}}{\rho_{ij}} + \frac{c_{Sij}}{\rho_{ij-1}} \right) (\beta^{(4)})_{ij}^{+}$$
$$= \left(\frac{c_{Sij}c_{Sij}}{\rho_{ij-1}} \gamma_{ij}^{(4)} + \frac{c_{Pij}c_{Sij-1}}{\rho_{ij}} \gamma_{ij}^{(2)} \right).$$
(7.70)

The total effect of vertical transverse propagation at the interface between cells $(i-1\,j)$ and (ij) is determined as follows

$$G_{ij}^{k} = \frac{\Delta t}{2\Delta x} \left(\mathcal{B}^{-}\mathcal{A}^{+}\Delta q_{ij}^{k} + \mathcal{B}^{+}\mathcal{A}^{+}\Delta q_{ij-1}^{k} \right.$$
$$\left. + \mathcal{B}^{+}\mathcal{A}^{-}\Delta q_{i+1\,j-1}^{k} + \mathcal{B}^{-}\mathcal{A}^{-}\Delta q_{i+1\,j}^{k} \right).$$
(7.71)

7.4.2 *Horizontal transverse propagation*

Analogously, horizontal transverse fluctuations $\mathcal{A}^{+}\mathcal{B}^{*}\Delta q_{ij}$ (the right-going transverse fluctuation), and $\mathcal{A}^{-}\mathcal{B}^{*}\Delta q_{ij}$ (the left-going transverse fluctuation) are determined by

$$\mathcal{A}^{-}\mathcal{B}^{-}\Delta q_{ij} = -c_{Pi-1\,j}(\alpha^{(1)})_{ij}^{-} r_{i-1\,j}^{(1)} - c_{Si-1\,j}(\alpha^{(3)})_{ij}^{-} r_{i-1\,j}^{(3)}, \qquad (7.72)$$

$$\mathcal{A}^{+}\mathcal{B}^{-}\Delta q_{ij} = c_{Pij}(\alpha^{(2)})_{ij}^{-} r_{ij}^{(2)} + c_{Sij}(\alpha^{(4)})_{ij}^{-} r_{ij}^{(4)}, \qquad (7.73)$$

$$\mathcal{A}^{-}\mathcal{B}^{+}\Delta q_{ij} = -c_{Pi-1\,j}(\alpha^{(1)})_{ij}^{+} r_{i-1\,j}^{(1)} - c_{Si-1\,j}(\alpha^{(3)})_{ij}^{+} r_{i-1\,j}^{(3)}, \qquad (7.74)$$

$$\mathcal{A}^{+}\mathcal{B}^{+}\Delta q_{ij} = c_{Pij}(\alpha^{(2)})_{ij}^{+} r_{ij}^{(2)} + c_{Sij}(\alpha^{(4)})_{ij}^{+} r_{ij}^{(4)}. \qquad (7.75)$$

Corresponding coefficients are given by

$$\left(\frac{c_{Pi-1\,j}}{\rho_{ij}} + \frac{c_{Pij}}{\rho_{i-1\,j}} \right) (\alpha^{(1)})_{ij}^{-}$$
$$= \left(-\frac{c_{Sij-1}c_{Pij}}{\rho_{ij-1}} \delta_{ij}^{(3)} - \frac{\lambda_{ij-1}}{\rho_{ij-1}\rho_{ij}} \delta_{ij}^{(1)} \right),$$
(7.76)

$$\left(\frac{c_{Pi-1\,j}}{\rho_{ij}} + \frac{c_{Pij}}{\rho_{i-1\,j}} \right) (\alpha^{(2)})_{ij}^{-}$$
$$= \left(-\frac{c_{Si\,j-1} c_{Pi-1\,j}}{\rho_{i\,j-1}} \delta_{ij}^{(3)} + \frac{\lambda_{i\,j-1}}{\rho_{i\,j-1} \rho_{i-1\,j}} \delta_{ij}^{(1)} \right), \tag{7.77}$$

$$\left(\frac{c_{Si-1\,j}}{\rho_{ij}} + \frac{c_{Sij}}{\rho_{i-1\,j}} \right) (\alpha^{(3)})_{ij}^{-}$$
$$= \left(-\frac{c_{Si\,j-1} c_{Si\,j-1}}{\rho_{ij}} \delta_{ij}^{(3)} - \frac{c_{Pi\,j-1} c_{Sij}}{\rho_{i\,j-1}} \delta_{ij}^{(1)} \right), \tag{7.78}$$

$$\left(\frac{c_{Si-1\,j}}{\rho_{ij}} + \frac{c_{Sij}}{\rho_{i-1\,j}} \right) (\alpha^{(4)})_{ij}^{-}$$
$$= \left(\frac{c_{Si\,j-1} c_{Si\,j-1}}{\rho_{i-1\,j}} \delta_{ij}^{(3)} - \frac{c_{Pi\,j-1} c_{Si-1\,j}}{\rho_{i\,j-1}} \delta_{ij}^{(1)} \right), \tag{7.79}$$

$$\left(\frac{c_{Pi-1\,j}}{\rho_{ij}} + \frac{c_{Pij}}{\rho_{i-1\,j}} \right) (\alpha^{(1)})_{ij}^{+}$$
$$= \left(\frac{c_{Sij} c_{Pij}}{\rho_{ij}} \delta_{ij}^{(4)} - \frac{\lambda_{ij}}{\rho_{ij} \rho_{ij}} \delta_{ij}^{(2)} \right), \tag{7.80}$$

$$\left(\frac{c_{Pi-1\,j}}{\rho_{ij}} + \frac{c_{Pij}}{\rho_{i-1\,j}} \right) (\alpha^{(2)})_{ij}^{+}$$
$$= \left(\frac{c_{Sij} c_{Pi-1\,j}}{\rho_{ij}} \delta_{ij}^{(4)} + \frac{\lambda_{ij}}{\rho_{i-1\,j} \rho_{ij}} \delta_{ij}^{(2)} \right), \tag{7.81}$$

$$\left(\frac{c_{Si-1\,j}}{\rho_{ij}} + \frac{c_{Sij}}{\rho_{i-1\,j}} \right) (\alpha^{(3)})_{ij}^{+}$$
$$= \left(-\frac{c_{Sij} c_{Si\,j-1}}{\rho_{ij}} \delta_{ij}^{(4)} + \frac{c_{Pij} c_{Sij}}{\rho_{ij}} \delta_{ij}^{(2)} \right), \tag{7.82}$$

$$\left(\frac{c_{Si-1\,j}}{\rho_{ij}} + \frac{c_{Sij}}{\rho_{i-1\,j}} \right) (\alpha^{(4)})_{ij}^{+}$$
$$= \left(\frac{c_{Sij} c_{Si\,j-1}}{\rho_{i-1\,j}} \delta_{ij}^{(4)} + \frac{c_{Pij} c_{Si-1\,j}}{\rho_{ij}} \delta_{ij}^{(2)} \right). \tag{7.83}$$

The total effect of horizontal transverse propagation at the interface between cells $(i\,j-1)$ and (ij) is determined as follows

$$F_{ij}^{k} = \frac{\Delta t}{2\Delta y} \left(\mathcal{A}^{-} \mathcal{B}^{-} \Delta q_{i\,j+1}^{k} + \mathcal{A}^{+} \mathcal{B}^{-} \Delta q_{i-1\,j+1}^{k} \right.$$
$$\left. + \mathcal{A}^{-} \mathcal{B}^{+} \Delta q_{ij}^{k} + \mathcal{A}^{+} \mathcal{B}^{+} \Delta q_{i-1\,j}^{k} \right). \tag{7.84}$$

The introduction of the transverse fluctuations improves the stability limit up to Courant number 1. Transverse fluctuations are included in the algorithm by

$$q_{ij}^{k+1} = q_{ij}^k + \Delta_{ij}^{up} - \frac{\Delta t}{\Delta x}(F_{i+1\,j}^k - F_{ij}^k) - \frac{\Delta t}{\Delta y}(G_{i\,j+1}^k - G_{ij}^k), \qquad (7.85)$$

where Δ_{ij}^{up} is the update for the first-order upwind Godunov method, of the form

$$\Delta_{ij}^{up} = -\frac{\Delta t}{\Delta x}(\mathcal{A}^-\Delta q_{i+1\,j} + \mathcal{A}^+\Delta q_{ij}) - \frac{\Delta t}{\Delta y}(\mathcal{B}^-\Delta q_{i\,j+1} + \mathcal{B}^+\Delta q_{ij}). \quad (7.86)$$

The whole algorithm is implemented in the form

$$q_{ij}^{k+1} = q_{ij}^k + \Delta_{ij}^{up} + \Delta_{ij}^{trans} - \frac{\Delta t}{\Delta x}(\bar{F}_{i+1\,j}^k - \bar{F}_{ij}^k) - \frac{\Delta t}{\Delta y}(\bar{G}_{i\,j+1}^k - \bar{G}_{ij}^k), \quad (7.87)$$

where Δ_{ij}^{up} is the update for the first-order upwind Godunov method, Δ_{ij}^{trans} represents the effect of transverse fluctuations are given by

$$\Delta_{ij}^{trans} = -\frac{\Delta t}{\Delta x}(F_{i+1\,j}^k - F_{ij}^k) - \frac{\Delta t}{\Delta y}(G_{i\,j+1}^k - G_{ij}^k). \qquad (7.88)$$

It is well known that the Lax-Wendroff scheme produces oscillations behind discontinuities [Liska and Wendroff (1998)]. The usual way to reduce spurious oscillations is to introduce limiter functions to modify the second-order corrections near discontinuities [LeVeque (2002a)]. Here, instead of limiters, the recent idea of using filters that are consistent with differential equations [Liska and Wendroff (1998)] is applied.

The corresponding composite scheme is obtained by application of the Godunov step after each three second-order Lax-Wendroff steps.

7.4.3 *Boundary conditions*

Boundary conditions are specified in terms of bulk quantities for boundary elements. For stress-free boundaries, the bulk stresses are set equal to zero. The excess quantities are not specified at the outer boundaries.

7.5 Numerical tests

A number of simulations has been performed to estimate the correctness and capabilities of the algorithm. They include, in particular, the following:

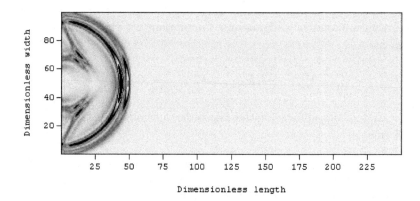

Fig. 7.1 Contour plot of elastic wavefronts from a point source in a homogeneous medium at 50 time steps.

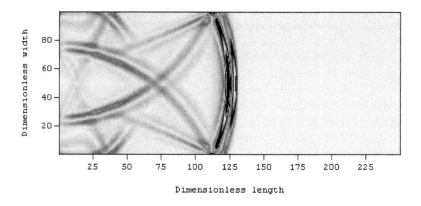

Fig. 7.2 Contour plot of elastic wavefronts from a point source in a homogeneous medium at 150 time steps.

• Elastic wavefronts from a point source at the boundary of a homogeneous medium;
• Elastic wave propagation in a layered medium;
• Elastic wave in a medium with periodically distributed inclusions.

All calculations are performed by means of the composite Lax-Wendroff-Godunov scheme with the Courant number equal to 1. Waves are excited by the prescription of non-zero normal component of the stress tensor within a finite aperture at the boundary for a short time.

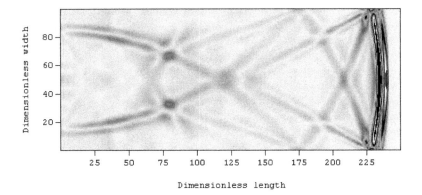

Fig. 7.3 Contour plot of elastic wavefronts from a point source in a homogeneous medium at 250 time steps.

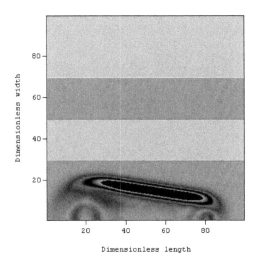

Fig. 7.4 Contour plot of elastic wave propagation in a periodic medium at 10 time steps.

As the first example, a short-time symmetrical excitation in the middle of a boundary of a homogeneous rectangular domain by sinusoidal normal stress is considered. All other boundaries are stress-free. Physical parameters for aluminium are used in the calculation: $c_p = 6420$ m/s, $c_s = 3040$ m/s, $\rho_0 = 2700$ kg/m^3.

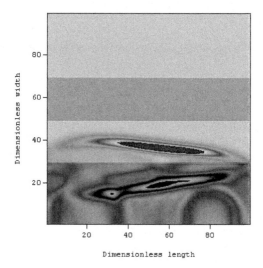

Fig. 7.5   Contour plot of elastic wave propagation in a periodic medium at 50 time steps.

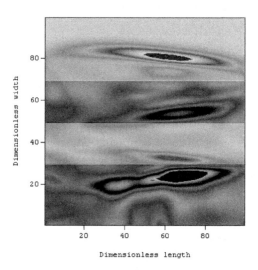

Fig. 7.6   Contour plot of elastic wave propagation in a periodic medium at 100 time steps.

Figures 7.1 - 7.3 represent snapshots of wavefronts after 50, 150, and 250 time steps, respectively, in terms of total displacement. These figures

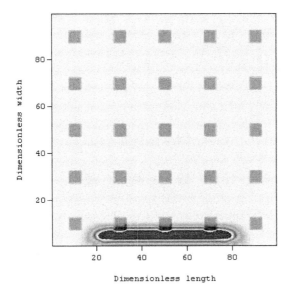

Fig. 7.7   Contour plot of elastic wave propagation in a periodic medium at 10 time steps.

demonstrate clearly how dilatational and shear waves propagate and interact with each other and with boundaries. It seems that the proposed algorithm gives a good correspondence with the almost classical situation.

Figures 7.4 - 7.6 show snapshots of the propagation of elastic wave, which is excited at a part of the bottom boundary with linear retardation, in a layered medium (corresponding density difference is also shown). In the inhomogeneous case, we need to prescribe additionally the physical parameters for copper: $c_p = 4560$ m/s, $c_s = 2600$ m/s, $\rho_0 = 8960$ kg/m^3. These figures clearly capture the reflection and transmission of the wave fronts at the interfaces. The pictures are asymmetric due to asymmetric loading at the bottom boundary.

The next three figures represent an elastic wave propagation in a medium (aluminium) with periodically distributed inclusions of another material (copper). The symmetric structure of wave fronts is formed due to the multiple reflections at the inclusions.

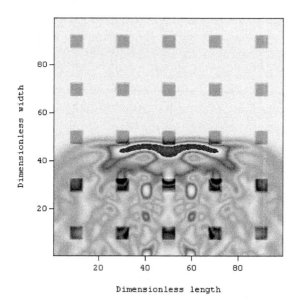

Fig. 7.8   Contour plot of elastic wave propagation in a periodic medium at 50 time steps.

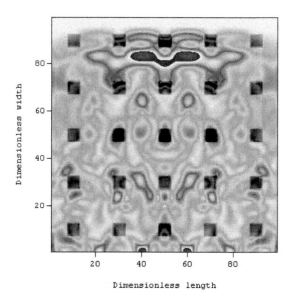

Fig. 7.9   Contour plot of elastic wave propagation in a periodic medium at 100 time steps.

7.6   Concluding remarks

The concept of excess quantities which was introduced for the thermodynamic description of the non-equilibrium state of elements following the thermodynamics of discrete systems is successfully used for the construction of a finite-volume numerical scheme for thermoelastic wave propagation in inhomogeneous media. The excess quantities appeared naturally in the integral balance laws for thermoelasticity; the laws serve as a basis for the formulation of the finite volume algorithm. At the same time, these quantities can be easily placed in correspondence with the fluctuations that appear in the formulation of the wave-propagation algorithm. This allows us to transform the wave-propagation algorithm into a thermodynamically consistent scheme using the idea of composition. The modification conserves all advantages of the wave-propagation algorithm, namely, high resolution, multidimensional motion, and stability up to a Courant number equal to unity.

The local equilibrium jump relations for adjacent elements provide the equivalence of the thermodynamic description at each level of refinement of the mesh. Within the composite wave-propagation algorithm, every discontinuity in parameters is taken into account by solving the Riemann problem at each interface between elements.

Chapter 8

Two-Dimensional Waves in
Functionally Graded Materials

In the one-dimensional example of wave propagation in FGM considered in chapter 5, the variation of properties of material was continuous. Usually, to solve elastic wave propagation in an FGM, the functionally graded medium is replaced by a multi-layered medium with fine homogeneous [Bruck (2000); Wang and Rokhlin (2004)] or inhomogeneous [Liu, Han and Lam (1999); Han, Liu and Lam (2001)] layers.

In the case of dynamic loads, different models of FGM are analyzed by Banks-Sills, Elasi and Berlin (2002). The analysis was carried out for the case in which the microstructure was modeled by embedded particles, in another case the gradation of the material was modeled by layers of different volume fraction, and in the last model the material properties were taken to vary continuously. It was concluded that all considered models produce more or less similar results.

We will apply a similar approach to analyze different FGM models in the case of propagation of two-dimensional stress waves in functionally graded materials under impact loading.

8.1 Impact loading of a plate

Here we consider the two-dimensional problem of the impulsive loading of a plate of thickness h and length $L \gg h$. The load is applied transversally at the central region of the plate upper surface of width a ($a < h$) (Fig. 8.1). The material of the plate is assumed to be compositionally graded along the thickness direction. The gradation is described in terms of the volume fraction of a ceramic reinforcing phase within a metal matrix. Following Li, Ramesh and Chin (2001), the three different timescales of the problem

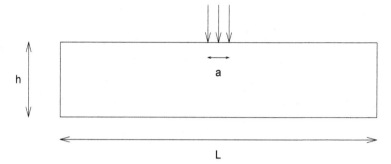

Fig. 8.1   The geometry of the problem.

can be introduced:

$$t_0 = \frac{h}{c_f}, \qquad t_b = \frac{L}{c_f}, \qquad t_r, \qquad (8.1)$$

where c_f is the speed of the fastest longitudinal wave, and t_r is the shortest rise time, corresponding to the applied loading. Since $L \gg h$, it follows that $t_0 \ll t_b$, so that the lateral boundaries of the plate are not actually involved in the calculations. As pointed out by Li, Ramesh and Chin (2001), if a volume fraction parameter alone is to be used to describe the gradation, then one must have $t_r \gg d_r/c_f$, where d_r is the reinforcement dimension.

We focus our attention on the case of dynamic loading of a plate where the wavelength of loading pulse is comparable with the plate thickness. This means that the wavelength is much larger than the size of inclusions and the distance of wave travel is relatively small.

Both metal and ceramics are assumed to behave as linear isotropic elastic media. In such a case, the governing equations of the problem can be represented in a form that is suitable for numerical simulations [Berezovski, Engelbrecht, and Maugin (2000); Berezovski and Maugin (2001)] - see Eqs. (7.1) - (7.2):

$$\rho_0(\mathbf{x})\frac{\partial v_i}{\partial t} = \frac{\partial \sigma_{ij}}{\partial x_j}, \qquad (8.2)$$

$$\frac{\partial \sigma_{ij}}{\partial t} = \lambda(\mathbf{x})\frac{\partial v_k}{\partial x_k}\delta_{ij} + \mu(\mathbf{x})\left(\frac{\partial v_i}{\partial x_j} + \frac{\partial v_j}{\partial x_i}\right), \qquad (8.3)$$

where, as previously, v_i are components of the velocity vector, σ_{ij} is the Cauchy stress tensor, ρ_0 is the density, λ and μ are the Lamé coefficients, δ_{ij} is the Kronecker delta. The indicated explicit dependence on the point

x means that the body is materially inhomogeneous in general, i.e. the properties of graded materials are reflected in $\rho(\mathbf{x})$, $\lambda(\mathbf{x})$, and $\mu(\mathbf{x})$.

Assuming that the plate is at rest for $t \leq 0$, system of Eqs. (8.2), (8.3) must be solved under the following initial conditions:

$$u_i(\mathbf{x}, 0) = 0, \qquad \sigma_{ij}(\mathbf{x}, 0) = 0. \tag{8.4}$$

The upper surface of the plate is subjected to a stress pulse given by

$$\sigma_{22}(x, 0, t) = \sigma_0 sin^2(\pi(t - 2t_r)/2t_r),$$
$$-a/2 < x < a/2, \quad 0 < t < 2t_r. \tag{8.5}$$

Other parts of upper and bottom surfaces are stress-free, and lateral boundaries are assumed to be fixed.

The problem (8.2)-(8.5) is solved numerically with distributions of material properties corresponding to the multi-layered model of metal-ceramic composite with averaged properties within layers and to the model of randomly embedded ceramic particles in a metal matrix with a prescribed volume fraction. The idea of direct numerical simulation of FGMs by means of regular [Banks-Sills, Elasi and Berlin (2002)] or random [Leggoe et al. (1998)] particle distribution is not a new one: it allows us to analyze the influence of shape and aspect ratio of particles on the FGM behaviour [Li and Ramesh (1998)]. Our goal is simply to show the differences in dynamical behavior of FGMs by using distinct models.

8.2 Material properties

In order to compute the overall strain/stress distributions in FGMs, one needs the appropriate estimates for properties of the graded layer, such as the Young's modulus, Poisson's ratio, etc. A large number of papers on the prediction of material properties of FGMs has been published (see, for example, [Gasik (1998); Cho and Oden (2000); Cho and Ha (2001)]). In most of these studies the averaging methods have been used which are simple and convenient to predict the overall thermomechanical response and properties. However, owing to the assumed simplifications, the validity of simplified models in real FGMs is affected by the corresponding detailed microstructure and other conditions. Still, the averaging methods may be selectively applied to FGMs subjected to both uniform and non-uniform overall loads with a reasonable degree of confidence.

Our main goal is to compare two distinct models of FGMs: (i) a multi-layered model of a metal-ceramic composite with averaged properties within

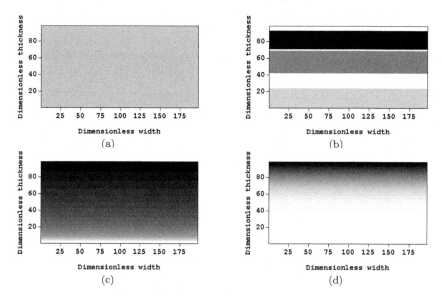

Fig. 8.2   Density distribution in metal-ceramic composite with ceramic reinforcement corresponding to the multilayered model with averaged properties within layers: a) uniform density, b) layered, c) continuously graded (high f front), d) continuously graded (low f front)(after [Li, Ramesh and Chin (2001)]).

each layer and (ii) a model of the same composite with randomly embedded ceramic particles in a metal matrix. For this purpose, it is sufficient to employ the linear rule of mixtures for the Young's modulus and Poisson's ratio of the graded layer in the multi-layered model of metal-ceramic composite with averaged properties within layers.

According to the linear rule of mixtures, the simplest estimate of any material property, a property $P(\mathbf{x})$ at a point $\mathbf{x}$ in two-phase metal-ceramic materials is approximated by a linear combination of volume fractions V_m and V_c and individual material properties of metal and ceramic constituents P_m and P_c:

$$P(\mathbf{x}) = P_m V_m(\mathbf{x}) + P_c V_c(\mathbf{x}). \tag{8.6}$$

The same individual material properties without any averaging are used in the model of the composite with randomly embedded ceramic particles in a metal matrix.

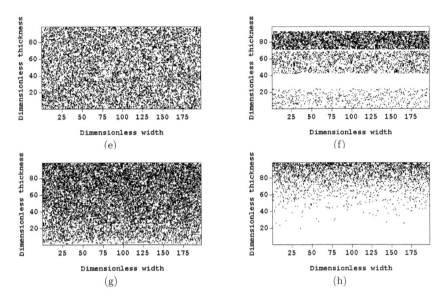

Fig. 8.3 Random particle distribution in metal-ceramic composite with ceramic reinforcement: e) uniform density, f) layered, g) graded (high f front), h) graded (low f front).

8.3 Numerical simulations

The above-described composite wave-propagation algorithm is applied to compare the models of discrete layers with averaging the material properties and of randomly embedded ceramic particles in a metal matrix.

To that purpose, four possible forms of the ceramic particulate reinforcement volume fraction variation along the thickness are considered in the case of the multi-layered model with averaged properties within layers: uniform, layered, and graded with two different distributions of volume fraction $f = V_c$ (Fig. 8.2), where the elastic properties of the metal matrix and ceramic reinforcement are the following [Li, Ramesh and Chin (2001)]: Young's modulus 70 GPa and 420 GPa, Poisson's ratio 0.3 and 0.17, and density 2800 kg/m^3 and 3100 kg/m^3, respectively. These structures are referred to as Cases A, B, C and D by Li, Ramesh and Chin (2001) in the axisymmetric case. Simultaneously, the same structures are represented by randomly embedded ceramic particles with the corresponding volume fraction (Fig. 8.3).

The same algorithm is applied for the numerical simulation of stress wave propagation in both models. In the computation, 98 elements are used in the thickness direction, so that the reinforcement dimension is as long as 250 μm. The rise time t_r for the loading was chosen as 0.75 μs. This satisfies the condition $t_r \gg d_r/c_f$, because the speed of the fastest longitudinal wave is equal to 11877 m/s corresponding to the data by Li, Ramesh and Chin (2001).

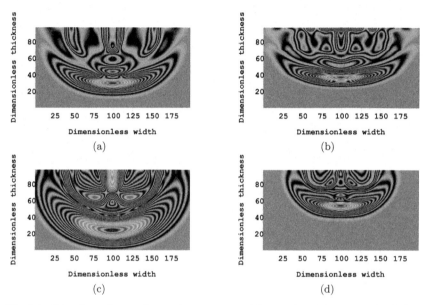

Fig. 8.4   Wavefronts in metal-ceramic composite with ceramic reinforcement (multilayered model) at 3 μs: a) uniform density, b) layered, c) continuously graded (high f front), d) continuously graded (low f front).

Typical examples of contour plots showing the full displacement fields are presented in Figs. 8.4 and 8.5 for the same density distribution in different models. Computations have been performed under the same conditions for distinct models of the same FGM: L=49 mm, h=24.5 mm, a=12.25 mm, σ_0=125 MPa.

The difference in the propagation speed of stress wave is clearly seen whereas the particle volume fraction was distributed similarly in both models. In addition, we have the distortion of the symmetrical shape of the wavefronts in the model with randomly embedded particles due to random particle distribution. The difference in the propagation speed of stress wave is clearly seen whereas the particle volume fraction was distributed similarly in both models. In addition, we have the distortion of the symmetrical shape of the wavefronts in the model with randomly embedded particles due to random particle distribution.

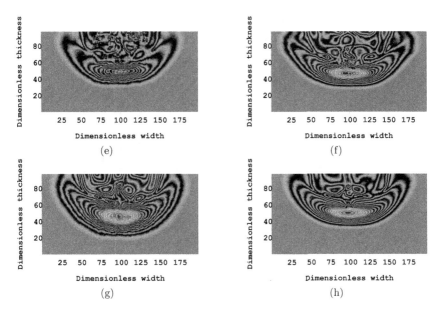

Fig. 8.5 Wavefronts in metal-ceramic composite with ceramic reinforcement (randomly embedded particles) at 3 μs: e) uniform density, f) layered, g) continuously graded (high f front), h) continuously graded (low f front).

One of the issues of interest in the use of layered or graded structures, for example, for armour applications was formulated by Li, Ramesh and Chin (2001) as follows: What is the effect of the layering or gradation on maxima of stresses and their distributions? This is important in order to optimize the structure relative to integrity under dynamic loading. In an attempt to answer the formulated question, let us consider the normal stress distribution along the centreline of the plate where maximal values of stress are expected.

8.4 Centreline stress distribution

The normal stress distribution along the centreline for the multi-layered model with averaged properties within layers is shown in Fig. 8.6. Since the results are given at the same instant of time, the difference in the position of normal stress profiles characterizes the corresponding difference in the

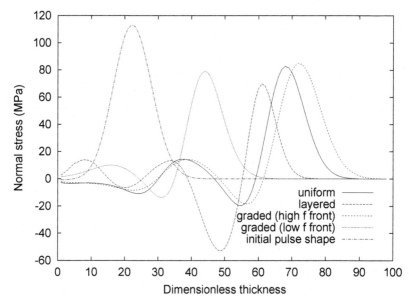

Fig. 8.6   Normal stress distribution along the centerline in metal-ceramic composite with ceramic reinforcement at 3 μs (multilayered model with averaged properties within layers).

speed of the stress wave in each structure in full accordance with the mean volume fraction of ceramic particles in each structure (0.360 for the uniform distribution, 0.226 for the layered case, 0.475 for the high f front, and 0.120 for the low f front). Obviously the alternating layering (case b) results in the decrease of the amplitude and of the transmitted wave compared with the homogenized case (a). The maximal tensile stresses are considerably higher for the layered case. Continuous grading yields the following (see Fig. 8.2). There is no significant differences in maximal amplitudes between the high and low f front grading (cases c, d) but there is a large difference in speeds. This effect is due to the material properties of the corresponding ceramic reinforcement.

The huge tensile stresses in the layered material due to the reflection through the structure dictates its rejection in order to provide the improvement of the integrity of the structure under dynamic loading.

The situation is slightly changed for the model of randomly embedded particles (Fig. 8.7). Here the difference in the speed of propagation of the stress waves in distinct structures is much less while the difference in their amplitudes became higher. Due to the high percentage of ceramic reinforce-

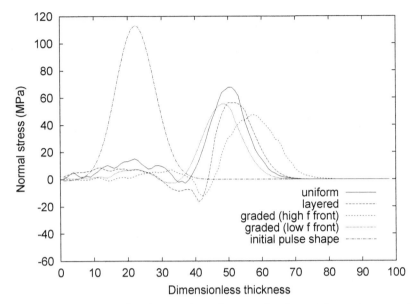

Fig. 8.7 Normal stress distribution along the centerline in metal-ceramic composite with ceramic reinforcement at 3 μs (model with randomly embedded particles).

ment in case c), the speed of the longitudinal wave keeps the highest value in all the calculated cases. The amplitude of the tensile stresses decreases significantly, especially for uniform and low f front grading (cases e) and h)). It seems that the latter structure (case h)) is the best choice for the damping of both the compressional and tensile stresses. This result is not so obvious in the case of the multilayered model with averaged properties within layers.

8.5   Wave interaction with functionally graded inclusion

Another example of differences in wave propagation in FGM represented by distinct models is given by the numerical simulation of wave interaction with a functionally graded inclusion. Density gradation in the circular inclusion is assumed to be Gaussian and represented by different models (Fig. 8.8):

- random distribution of embedded ceramic particles;
- multilayered medium;
- continuously changing density.

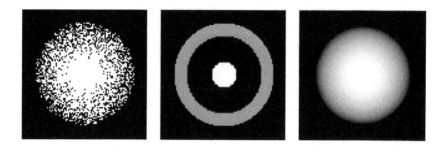

Fig. 8.8   Density gradation in inclusion represented by particles, layers, and continuously changing structure.

Material properties of pure metal (Al 6061-TO) and pure ceramics (TiC) are presented in Table 8.1 [Banks-Sills, Elasi and Berlin (2002)].

Table 8.1   Properties of materials.

Material	E (GPa)	ν	ρ (kg/m^3)
TiC	480	0.20	4920
Al 6061-TO	69	0.33	2700

In all cases a constant stress loading is applied at the right boundary. The left boundary is fixed. Other boundaries are stress-free (Fig. 8.9).

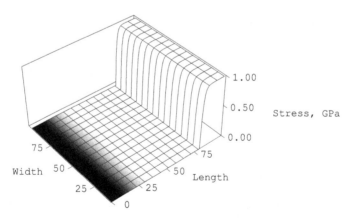

Fig. 8.9   Initial stress distribution.

It should be noted that in the homogeneous case the longitudinal stress distribution restores its shape after every four reflections from the left and right boundaries of the specimen.

Numerical simulations have been performed using the different FGM models by means of the finite volume algorithm described above. Results of simulations are represented in terms of the longitudinal stress distribution at different time instances.

Figure 8.10 shows the stress distribution at 1 μs computed by using different models of FGM.

As we can see, the stress distribution corresponding to the model of randomly embedded particles (upper part of the figure) differs from the stress distribution in the multilayered medium (middle part) and even more so from the continuously changing case (bottom part of the figure) at this time instant.

However, continuation of computations leads to more or less similar stress distributions for sufficiently large time duration. This is illustrated by Fig. 8.11. This means that different models of FGM may give similar results for steady-state (or quasi-static) situations, but the results can be essentially different for short time dynamics.

8.6   Concluding remarks

Theoretical prediction of dynamic behavior of FGMs depends on how well their properties are modeled in computer simulations. While many averaging models of the properties of FGMs are widely accepted, a more natural model of a matrix with randomly embedded particles is seldom used because of numerical difficulties in the case of rapidly-varying properties of the medium. This difficulty is overcome here by using the wave-propagation algorithm [LeVeque (1997)] and its modifications [Berezovski, Engelbrecht, and Maugin (2000); Berezovski and Maugin (2001)]. Within the composite wave-propagation algorithm, every discontinuity in parameters is taken into account by solving the Riemann problem at each interface between discrete elements. The reflection and transmission of waves at each interface are handled automatically for the considered inhomogeneous media.

The results of performed numerical simulations of stress wave propagation in FGMs show a significant difference between characteristics of wave fields in the models of discrete layers with averaging the material properties and of randomly embedded ceramic particles in a metal matrix. This means that a model of FGM without averaging of material properties can give a more detailed information about the dynamic behavior of a chosen structure, which may be used in its optimization for a particular situation.

a)

b)

c)

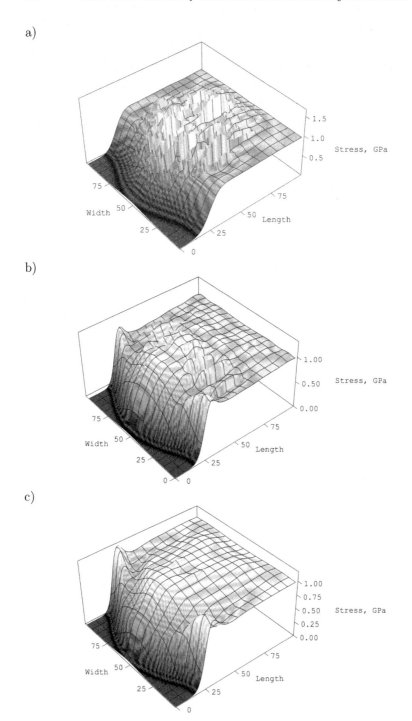

Fig. 8.10   Stress distribution at 1 μs, a) random particles, b) layered, c) continuous gradation.

a)

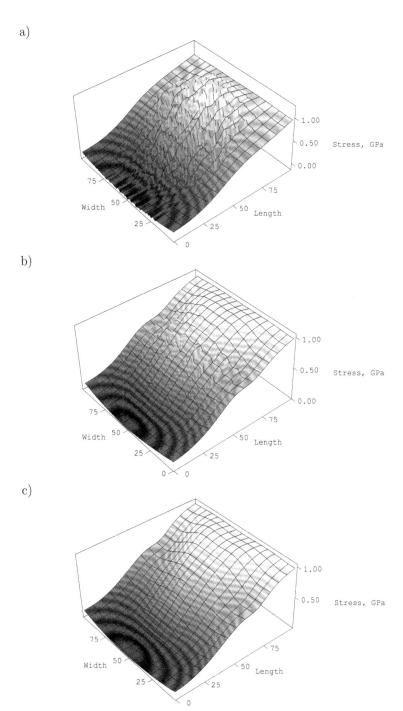

b)

c)

Fig. 8.11   Stress distribution at 5 μs, a) random particles, b) layered, c) continuous gradation.

It should be noted that the size, the shape, the clustering, and inhomo-geneities in the random distribution of embedded reinforcement particles may affect the results of simulation in each particular case. Nevertheless, at least some of the mentioned effects can be easily taken into account by means of the developed numerical scheme.

Chapter 9

Phase Transitions Fronts in Two Dimensions

Here we extend the one-dimensional considerations concerning phase-transition front propagation (see chapter 6) to a two-dimensional case. The main question here is how to describe the kinematics of the phase boundary.

9.1 Material velocity at the phase boundary

We recall the results from chapter 6. The material velocity at the phase boundary can be determined by means of jump relation for linear momentum (4.15)

$$\bar{V}_N[\rho_0\bar{v}_i] + N_j[\bar{\sigma}_{ij}] = 0. \tag{9.1}$$

Using the kinematic relationship between material and physical velocities [Maugin (1995)]

$$\mathbf{v} = -\mathbf{F} \cdot \mathbf{V}, \tag{9.2}$$

we can represent the jump relation for the linear momentum (9.1) in terms of components

$$-\bar{V}_i\langle\rho_0\rangle \left[\frac{\partial\bar{u}_k}{\partial x_j}\right]\langle\bar{V}_j\rangle + [\bar{\sigma}_{ik}] = 0. \tag{9.3}$$

In the two-dimensional case, four equations yield

$$-\bar{V}_1\langle\rho_0\rangle \left(\left[\frac{\partial\bar{u}_1}{\partial x_1}\right]\langle\bar{V}_1\rangle + \left[\frac{\partial\bar{u}_1}{\partial x_2}\right]\langle\bar{V}_2\rangle \right) + [\bar{\sigma}_{11}] = 0, \tag{9.4}$$

$$-\bar{V}_2\langle\rho_0\rangle \left(\left[\frac{\partial\bar{u}_1}{\partial x_1}\right]\langle\bar{V}_1\rangle + \left[\frac{\partial\bar{u}_1}{\partial x_2}\right]\langle\bar{V}_2\rangle \right) + [\bar{\sigma}_{21}] = 0, \tag{9.5}$$

161

$$-\bar{V}_1 \langle \rho_0 \rangle \left(\left[\frac{\partial \bar{u}_2}{\partial x_1} \right] \langle \bar{V}_1 \rangle + \left[\frac{\partial \bar{u}_2}{\partial x_2} \right] \langle \bar{V}_2 \rangle \right) + [\bar{\sigma}_{12}] = 0, \qquad (9.6)$$

$$-\bar{V}_2 \langle \rho_0 \rangle \left(\left[\frac{\partial \bar{u}_2}{\partial x_1} \right] \langle \bar{V}_1 \rangle + \left[\frac{\partial \bar{u}_2}{\partial x_2} \right] \langle \bar{V}_2 \rangle \right) + [\bar{\sigma}_{22}] = 0. \qquad (9.7)$$

The sum of the first and fourth of these equations gives

$$-\langle \rho_0 \rangle \left([\bar{\varepsilon}_{11}] \bar{V}_1^2 + 2 [\bar{\varepsilon}_{12}] \bar{V}_1 \bar{V}_2 + [\bar{\varepsilon}_{22}] \bar{V}_2^2 \right) + [\bar{\sigma}_{11} + \bar{\sigma}_{22}] = 0, \qquad (9.8)$$

whereas from the second and third equations we have

$$-\langle \rho_0 \rangle \bar{V}_1 \bar{V}_2 \left([\bar{\varepsilon}_{11}] \bar{V}_1^2 + 2 [\bar{\varepsilon}_{12}] \bar{V}_1 \bar{V}_2 + [\bar{\varepsilon}_{22}] \bar{V}_2^2 \right) + (\bar{V}_1^2 + \bar{V}_2^2)[\bar{\sigma}_{12}] = 0. \qquad (9.9)$$

It follows from Eqs. (9.8) and (9.9) that

$$\bar{V}_1 \bar{V}_2 [\bar{\sigma}_{11} + \bar{\sigma}_{22}] = (\bar{V}_1^2 + \bar{V}_2^2)[\bar{\sigma}_{12}]. \qquad (9.10)$$

Similar consideration with subtraction of corresponding equations provides the equation

$$\bar{V}_1 \bar{V}_2 [\bar{\sigma}_{11} - \bar{\sigma}_{22}] = (\bar{V}_1^2 - \bar{V}_2^2)[\bar{\sigma}_{12}], \qquad (9.11)$$

which can be reduced to a simpler relation

$$\bar{V}_1^2 [\bar{\sigma}_{22}] = \bar{V}_2^2 [\bar{\sigma}_{11}]. \qquad (9.12)$$

Substituting relations (9.10) and (9.12) into Eq. (9.8), we can determine the components of the material velocity at the phase boundary as

$$\langle \rho_0 \rangle \bar{V}_1^2 = \frac{[\bar{\sigma}_{11} + \bar{\sigma}_{22}][\bar{\sigma}_{11}]}{[\bar{\varepsilon}_{11}][\bar{\sigma}_{11}] + 2[\bar{\varepsilon}_{12}][\bar{\sigma}_{12}] + [\bar{\varepsilon}_{22}][\bar{\sigma}_{22}]}, \qquad (9.13)$$

$$\langle \rho_0 \rangle \bar{V}_2^2 = \frac{[\bar{\sigma}_{11} + \bar{\sigma}_{22}][\bar{\sigma}_{22}]}{[\bar{\varepsilon}_{11}][\bar{\sigma}_{11}] + 2[\bar{\varepsilon}_{12}][\bar{\sigma}_{12}] + [\bar{\varepsilon}_{22}][\bar{\sigma}_{22}]}. \qquad (9.14)$$

Just these expressions for the velocity components has been used in the numerical simulations.

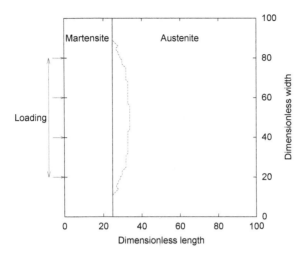

Fig. 9.1   Phase-transition front propagation under non-plane wave impact: solid line - initial phase boundary position, dashed line - phase boundary position after interaction, grid: 100 × 100 elements.

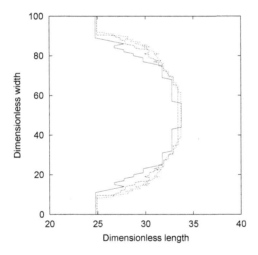

Fig. 9.2   Phase-transition front position under non-plane wave impact at 100 time steps: solid line - 100 × 100 elements grid, dashed line - 200 × 200 elements grid, dotted line - 400 × 400 elements grid.

9.2   Numerical procedure

The applied numerical scheme has been described in detail in chapter 7. This is the wave-propagation algorithm in terms of excess quantities.

The use of excess quantities allows us to extend the applicability of the wave-propagation algorithm to the computations of moving discontinuities. Namely, at the phase boundary, the continuity of normal components of excess quantities is applied, as it was already done in the one-dimensional case. Then the driving forces are computed in vertical and horizontal directions. If the value of the driving force overcomes its critical value, a virtual displacement of the front is calculated by means of the velocities of the front, indicated in the previous section. If the (algebraic) sum of these virtual displacements is larger than the size of the computational cell, the state of the latter is changed from austenite to martensite. Remaining part of calculations is the same as in the case of wave propagation.

9.3   Interaction of a non-plane wave with phase boundary

The first example of the two-dimensional simulations is similar to that considered in chapter 6. The geometry of the problem is shown in Fig. 9.1. The pulse loading is applied at a part of the left boundary of the computation domain by prescribing a time variation of a component of the stress tensor. The problem becomes two-dimensional. Upper and bottom boundaries are stress-free, the right boundary is assumed to be rigid. The time-history of loading is shown in Fig. 6.7.

Material properties correspond to Cu-14.44Al-4.19Ni shape-memory alloy [Escobar and Clifton (1993)] in austenitic phase: the density $\rho = 7100$ kg/m^3, the elastic modulus $E = 120$ GPa, the shear wave velocity $c_s = 1187$ m/s, the dilatation coefficient $\alpha = 6.75 \cdot 10^{-6}$ 1/K. For the martensitic phase we choose, respectively, $E = 60$ GPa, $c_s = 1055$ m/s, with the same density and dilatation coefficient as above.

As a result of interaction between the stress pulse and the phase boundary, the initially straight phase boundary is deformed to the form shown in Fig. 9.1. At this stage, we also compared the results of computation of the same problem by refining the mesh. The corresponding results are shown in Fig. 9.2. Here the longitudinal coordinate is zoomed to exhibit the differences more clearly. As one can see, the refining of the mesh leads to a finer description of the phase-transition front, but it does not change the overall shape significantly.

The next example concerns the behavior of the phase-transition front under increasing of the loading magnitude. Corresponding results are shown in Fig. 9.3, where two distinct final shapes of the phase-transition

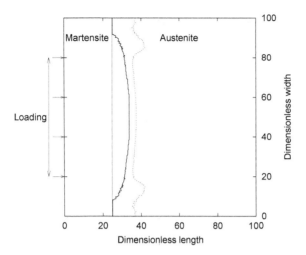

Fig. 9.3 Phase transition front at 100 time steps: straight line - initial position, solid line corresponds to the maximum magnitude of the loading 0.7 GPa, dotted line - 1.4 GPa (400 × 400 elements grid).

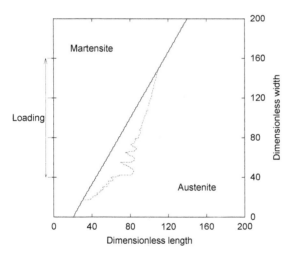

Fig. 9.4 Oblique impact of the stress wave: solid line - initial position of the phase boundary, dashed line - phase-transition front at 100 time steps.

front relate to magnitudes of loading equal to 0.7 and 1.4 GPa, respectively.

Figure 9.4 shows the phase-transition front position in the case of an oblique impact of the stress wave on the phase boundary. The result ex-

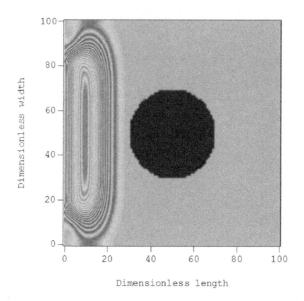

Fig. 9.5   Contour plot of stress distribution at 30 time steps.

hibited is tantamount to the formation of fingers or dendrites.

9.4   Wave interaction with martensitic inclusion

The same numerical procedure is applied to the simulation of the wave interaction with a martensitic inclusion in an austenite environment. Material properties and loading conditions are the same as in previous example. Initial shape of the martensitic inclusion is shown in Fig. 9.5.

Next figure shows the interaction of the pulse with the martensitic inclusion. Due to the difference in material properties of martensite and austenite, the velocity of the wave in martensite is less than this in austenite. Simultaneously, the phase transformation process is induced at the boundary between martensite and austenite. The final stage of the interaction is presented in Fig. 9.7 in terms of stress distribution. Changed shape of the martensitic inclusion after the interaction with the stress pulse is shown in Fig. 9.8. One can see the growth of the martensitic inclusion in all directions, but the growth in the rear side (right part of Fig. 9.8) is more than that in the front side (left part of Fig. 9.8).

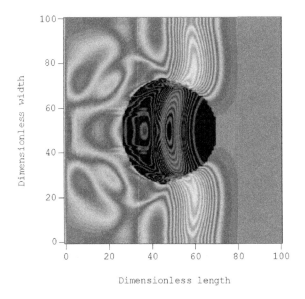

Fig. 9.6   Contour plot of stress distribution at 80 time steps.

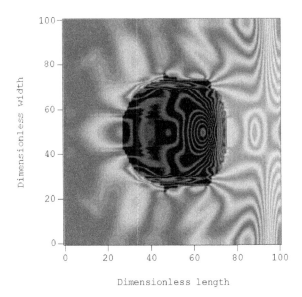

Fig. 9.7   Contour plot of stress distribution at 120 time steps.

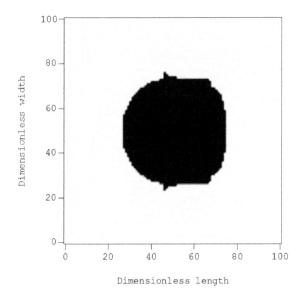

Fig. 9.8   Inclusion shape after interaction.

9.5   Concluding remarks

Rapid variations in the properties of phases require at least a second-order accuracy of the relevant algorithms. Finite-volume numerical methods are based on the integration of governing equations over a control volume, which includes a grid element and a time step. This means that the resulting numerical scheme is expressed in terms of averaged field variables and averaged fluxes at boundaries of the grid elements. The equations of state determining the properties of a medium are also assumed to be valid for the averaged quantities. In fact, this is an assumption of local equilibrium inside the grid element, where the local equilibrium state is determined by the averaged values of field variables. To obtain a high-order accuracy, the step-wise distribution of the field variables is changed to a piece-wise linear (or even non-linear) distribution over the grid. Such a reconstruction leads to a better approximation from the mathematical point of view and provides a high-order accuracy together with a certain procedure for suppressing spurious oscillations during computation. However, from the thermodynamic point of view, the reconstruction destroys the local equilibrium inside grid cells. This means that the equations of state are not valid in this case and even the meaning of thermodynamic variables (e.g. temperature and entropy) is questionable. A possible solution of this problem is the description of the non-equilibrium states inside the grid elements in the framework of the thermodynamics of discrete systems. Excess quantities can be introduced to the finite-volume schemes in a natural way. As a consequence, a thermodynamically consistent form of the finite-volume numerical method for the simulation of phase-transition front propagation is obtained.

Chapter 10

Dynamics of a Straight Brittle Crack

Crack propagation is another and more often observed example of moving discontinuity. As distinct from a two-dimensional boundary between phases, the crack front is a one-dimensional line.

In the framework of the linear elasticity theory [Freund (1990)], the tip of a moving crack is the sole energy sink for the elastic energy that is released from the surrounding material as fracture occurs. The crack tip equation of motion follows from the balance between the dynamic energy release rate, G, and the specific fracture energy, Γ, which is the amount of energy needed to create a crack of unit area. The equation for the energy balance at the crack tip is predicted to have the form [Freund (1990)]:

$$\Gamma \approx G(l)\left(1 - \frac{V_C}{c_R}\right),\tag{10.1}$$

where l is the instantaneous crack length, V_C is the crack's velocity, c_R is the Rayleigh wave speed, and $G(l)$ is the amount of energy per unit area present at the tip of a static crack of length l and it contains all of the effects of the applied stresses and specimen geometry. The equation of motion for a moving crack follows from the inversion of Eq. (10.1):

$$V_C(l) = c_R\left(1 - \frac{\Gamma}{G(l)}\right).\tag{10.2}$$

Relation (10.2) is used in numerical calculations of crack propagation (e.g., [Réthoré, Gravouil and Combescure (2004)]). However, Eq. (10.2) predicts that V_C can be arbitrarily close to its limiting velocity, c_R, as G can be arbitrarily large. At the same time, experimentally observed limiting crack speeds are significantly smaller than the Rayleigh wave velocity even in nominally brittle materials [Fineberg and Marder (1999); Ravi-Chandar (2004)]. Though this difference is explained via dynamic crack instability

[Fineberg and Marder (1999); Sharon and Fineberg (1999); Ravi-Chandar (2004)], the equation of motion for a crack still remains under question. As it was discussed in chapter 2, the crack front can be considered as a singular set of material points in the continuum theory similarly to phase-transition fronts [Maugin (2000)]. This means that the local equilibrium jump relations derived in chapter 3 can also be applied at the crack front. As distinct from the phase transition front problem, the problem of crack motion cannot be reduced to a one-dimensional formulation, because the driving force is calculated by means of the J-integral. At the same time, the jump relations at the crack front are simpler than in the case of moving phase boundary, since the state behind the crack front has completely zero values of any parameter. Therefore, the values of jumps coincide with the values of the corresponding (local equilibrium) quantities in front of the crack front.

In what follows we describe how the general relationship between driving force and the velocity of discontinuity derived in [Berezovski and Maugin (2005b)] can be specialized for the straight-through mode I crack.

10.1   Formulation of the problem

The simplest formulation of the crack propagation problem corresponds to mode I fracture in thin plates. We consider the crack propagation in a thin cracked plate subjected to a load as shown in Fig. 10.1.

In the framework of linear elasticity (cf. chapter 4), we can write the bulk equations in a homogeneous isotropic body in the absence of body force as follows:

$$\rho_0 \frac{\partial v_i}{\partial t} = \frac{\partial \sigma_{ij}}{\partial x_j}, \tag{10.3}$$

$$\frac{\partial \sigma_{ij}}{\partial t} = \lambda \frac{\partial v_k}{\partial x_k} \delta_{ij} + \mu \left(\frac{\partial v_i}{\partial x_j} + \frac{\partial v_j}{\partial x_i} \right), \tag{10.4}$$

where t is time, x_j are spatial coordinates, v_i are components of the velocity vector, σ_{ij} is the Cauchy stress tensor, ρ_0 is the density, λ and μ are the Lamé coefficients.

In the case of thin plates, the problem can be simplified by means of either plane stress approximation (thin strip geometry, $\sigma_{i3} = 0, i = 1, 2, 3$) or plane strain approximation (thin plate geometry, displacement components $u_{i3} = 0, i = 1, 2, 3$). These approximations lead to the formulations of the crack propagation problem, which are indistinguishable from each other except for the difference in values of dilatational wave speed. Corresponding solutions can be found elsewhere (cf., [Freund (1990); Ravi-Chandar (2004)]). These solutions should be in correspondence with the description of the problems in material formulation of continuum mechanics.

10.2 Stationary crack under impact load

First, we check the suitability of the wave-propagation algorithm for the solution of the problem of a body with a stationary crack subjected to impact load. The computations are carried out for an edge cracked specimen shown in Fig. 10.1. A state of plane strain is adopted. The specimen di-

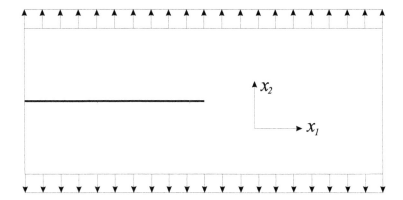

Fig. 10.1   Model problem for a crack in a plate.

mensions are taken to be $b = 20.0$ mm and $h = 4.0$ mm. Initial crack length is $a = 10$ mm. These dimensions ensure that no unloading waves from the lateral boundary of the specimen reach the crack tip during the period of interest.

The problem is formulated in the same way as in [Duarte et al. (2001)]. Lateral boundaries and crack faces are traction-free. A plane tensile pulse of magnitude $\sigma^* = 63750$ Pa propagating towards the crack plane from the upper and bottom surfaces of the specimen is prescribed at $t > 0$. Material properties correspond to a linear elastic material with $E = 2.0 \times 10^{11}$ Pa, $\nu = 0.3$, and $\rho = 7833.0$ kg/m^3. A uniform rectangular mesh is used for calculations. The mesh has 250 and 100 elements in the x_1- and x_2-directions, respectively. Computations are performed with the Courant number equal to 1.

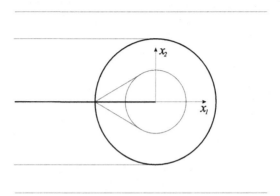

Fig. 10.2   Wave-fronts for diffraction of a plane wave by a semi-infinite crack. Theoretical prediction.

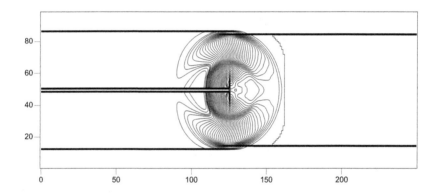

Fig. 10.3   Wave-fronts for diffraction of a plane wave by a stationary crack. Numerical simulation.

Figure 10.2 shows the scattered fields radiating out from the crack tip in the case of a semi-infinite crack under plane wave loading [Freund (1990)]. The leading wave-front is a cylindrically diffracted longitudinal wave, of radius $c_p\,t$, centered at the crack tip. Due to the coupling of dilatational and shear waves, the scattered field includes a cylindrical shear wave-front, of radius $c_s\,t$, centered at the crack tip, and the associated plane fronted headwaves travelling at speed c_s. Here c_p and c_s are the longitudinal wave speed and the shear wave speed, respectively.

Similar wave-fronts can be seen in Fig. 10.3, where the snapshot of full displacement field after 85 time steps is shown; this has been computed by means of the wave-propagation algorithm. It seems that the conservative wave-propagation algorithm represents the wave fronts as well.

10.3 Jump relations at the crack front

To be able to compute the crack motion, we need to predict the crack tip velocity. The crack front in the considered thin plate problem is a straight line in the $x_1 - x_3$ plane (x_3 is orthogonal to the plane of Fig. 10.1). In the plane the crack front line may be seen as a discontinuity between matter (in front of the crack) and vacuum (in the crack). Accordingly, we envision the exploitation of the local equilibrium jump relations established for a moving material discontinuity surface in chapter 3, but adapt these to the present situation.

The jump relation across crack front C corresponding to the balance of linear momentum (Eq. (4.15)) reads, in general,

$$V_C[\rho_0 v_i] + N_j[\sigma_{ij}] = 0. \tag{10.5}$$

Here V_C is the material velocity of the crack front along normal N_j. The jump relations across the crack front C corresponding to the balance of canonical momentum and entropy inequality should exhibit source terms [Maugin (1997)]:

$$V_C[\mathcal{P}_i] + N_j[b_{ij}] = -f_i, \tag{10.6}$$

$$V_C[S] - N_j[Q_j/\theta] = \sigma_C \geq 0, \tag{10.7}$$

where $\mathcal{P}_i$ is the pseudo-momentum in the small-strain approximation, b_{ij} is the dynamical Eshelby stress tensor, Q_j is the material heat flux, S is the entropy per unit volume, θ is temperature, σ_C is unknown scalar, and f_i is an unknown material force. The latter quantities are constrained to satisfy the second law of thermodynamics at the crack front C such that [Maugin (1997)]

$$f_i V_i = \theta_C \sigma_C \geq 0. \tag{10.8}$$

Due to non-zero entropy production at the moving crack front, we need to take into account the temperature dependence of the free energy even if we consider the isothermal case. Isothermal conditions are used due to the correspondence to linear elasticity theory. Adiabatic conditions may be more appropriate, but both in isothermal and in adiabatic cases temperature field is determined independently from other fields, and the isothermal case is much simpler to handle.

10.4   Velocity of the crack in mode I

Suppose that stress and strain fields are determined. We can estimate the velocity of the crack by means of the jump relation for linear momentum (10.5):

$$V_C[\rho_0 v_i] + N_j[\bar{\sigma}_{ij} + \Sigma_{ij}] = 0, \tag{10.9}$$

where overbars denote local equilibrium values, square brackets denote jumps, N_j is the normal to the crack front, and the excess stress Σ_{ij} is introduced. In the small strain approximation, the material velocity V_j is connected with the physical velocity v_i by [Maugin (1993)]

$$v_i = -(\delta_{ij} + \frac{\partial \bar{u}_i}{\partial x_j})V_j. \tag{10.10}$$

Inserting the latter relation into the former one, we have

$$V_C \left[\rho_0(\delta_{ij} + \frac{\partial \bar{u}_i}{\partial x_j})V_j \right] - N_j[\bar{\sigma}_{ij} + \Sigma_{ij}] = 0. \tag{10.11}$$

Remember that the crack front in the thin plate problem is a straight line in $x_1 - x_3$ plane, propagating in the x_1 direction. The projection on the normal to the crack front reduces the last expression to

$$V_C \left[\rho_0(1 + \bar{\varepsilon}_{11})V_1 \right] - [\bar{\sigma}_{11} + \Sigma_{11}] = 0, \tag{10.12}$$

where $\bar{\sigma}_{11}$ is the component of the stress tensor normal to the crack front. Since we have no material behind the crack front, jumps are equal to values of quantities in front of the crack front which leads to

$$V_C^2 = \frac{\bar{\sigma}_{11} + \Sigma_{11}}{\rho_0(1 + \bar{\varepsilon}_{11})}. \tag{10.13}$$

However, we are unable to determine the exact values of the stress components at the crack tip due to the square-root singularity. Therefore, we apply the local equilibrium jump relation (6.32) as in the case of the phase transition front [Berezovski and Maugin (2004)]:

$$\left[\bar{\sigma}_{11} + \bar{\theta} \left(\frac{\partial S}{\partial \varepsilon_{11}} \right)_{\sigma} + \Sigma_{11} \right] = 0. \tag{10.14}$$

It follows from Eq. (10.14) that the values of the normal stresses at the crack front are determined by the entropy derivative at the crack front

$$\bar{\sigma}_{11} + \Sigma_{11} = -\bar{\theta} \left(\frac{\partial S}{\partial \varepsilon_{11}} \right)_{\sigma}. \tag{10.15}$$

As follows from the entropy jump at a discontinuity (10.7) and the expression for entropy production (10.8), in the isothermal case entropy at the crack front is dependent only on the driving force

$$S = \frac{f_C}{\bar{\theta}}. \tag{10.16}$$

Therefore, for the entropy derivative on the right-hand side of Eq. (10.15) we obtain

$$\bar{\sigma}_{11} + \Sigma_{11} = -\bar{\theta} \left(\frac{\partial S}{\partial \varepsilon_{11}} \right)_{\sigma} = \frac{f_C}{\bar{\theta}} \left(\frac{\partial \bar{\theta}}{\partial \varepsilon_{11}} \right)_{\sigma} - \left(\frac{\partial f_C}{\partial \varepsilon_{11}} \right)_{\sigma}. \tag{10.17}$$

In the isothermal case, the second derivative on the right-hand side of the last relation can be neglected because stresses and strains are fixed simultaneously. This means that the normal stresses at the crack front can be expressed in terms of the driving force

$$\bar{\sigma}_{11} + \Sigma_{11} = \frac{f_C}{\bar{\theta}} \left(\frac{\partial \bar{\theta}}{\partial \varepsilon_{11}} \right)_{\sigma} = f_C \frac{2(\lambda + \mu)}{\bar{\theta} \alpha (3\lambda + 2\mu)} = A f_C, \tag{10.18}$$

where α is the thermal expansion coefficient.

It is well known that the driving force acting at the crack tip is the energy release rate, which can be calculated by means of the celebrated J-integral ([Maugin (2000)], e.g.). The dynamic J-integral for a homogeneous cracked body [Atkinson and Eshelby (1968); Kostrov and Nikitin (1970); Freund (1972); Maugin (1994)] can be expressed in the case of mode I straight crack as follows:

$$J = \lim_{\Gamma \to 0} \int_{\Gamma} \left((W + K)\delta_{1j} - \sigma_{ij} \frac{\partial u_i}{\partial x_1} \right) n_j d\Gamma. \tag{10.19}$$

Here n_j is the unit vector normal to an arbitrary contour Γ pointing outward of the enclosed domain. The kinetic energy density, K, is given by

$$K = \frac{1}{2} \rho_0 \mathbf{v}^2. \tag{10.20}$$

In the two-dimensional case, the driving force is related to the value of the J-integral (10.19) as follows:

$$f_C = \frac{J}{l}, \tag{10.21}$$

where l is a scaling factor which has dimension of length.

Summing up, we can represent stresses at the crack front as

$$\bar{\sigma}_{11} + \Sigma_{11} = \frac{AJ}{l}. \tag{10.22}$$

Here we should make a supposition concerning the stress excess Σ_{11}.

10.4.1   *Zero excess stress*

The simplest estimate is found by assuming that the value of the excess stress Σ_{11} is zero at the crack front. This means that Eq. (10.13) is reduced to

$$V_C^2 = \frac{\bar{\sigma}_{11}}{\rho_0(1 + \bar{\varepsilon}_{11})}. \tag{10.23}$$

In the framework of the linear theory, we can expect a linear stress-strain relation between the local equilibrium stress, $\bar{\sigma}_{11}$, and the local equilibrium strain, $\bar{\varepsilon}_{11}$,

$$\bar{\sigma}_{11} = B\bar{\varepsilon}_{11}. \tag{10.24}$$

Inserting Eq. (10.24) into Eq. (10.23), we have

$$V_C^2 = \frac{\bar{\sigma}_{11}}{\rho_0\left(1 + \bar{\sigma}_{11}/B\right)}. \tag{10.25}$$

The value of the coefficient B is determined from the condition that the velocity of the crack front should approach the Rayleigh wave velocity c_R at high values of $\bar{\sigma}_{11}$. It follows from Eq. (10.25) that

$$\lim_{\sigma \to \infty} V_C^2 = c_R^2 = \frac{B}{\rho_0}. \tag{10.26}$$

Taking into account relations (10.22) and (10.26), we can represent Eq. (10.25) as follows

$$\frac{V_C^2}{c_R^2} = \left(1 + \frac{\rho_0 c_R^2 l}{AJ}\right)^{-1}. \tag{10.27}$$

This means that for sufficiently small values of $\rho_0 c_R^2 l / AJ$

$$\frac{V_C^2}{c_R^2} \approx 1 - \frac{\rho_0 c_R^2 l}{AJ}. \tag{10.28}$$

Extracting the root from both sides of the last equation, we obtain

$$\frac{V_C}{c_R} \approx \sqrt{1 - \frac{\rho_0 c_R^2 l}{AJ}} \approx 1 - \frac{\rho_0 c_R^2 l}{2AJ}. \tag{10.29}$$

Thus, assuming zero excess stress at the crack front, we come to the well-known relation for the crack velocity (10.2).

10.4.2 *Non-zero excess stress*

Certainly, there are other possibilities in the choice of the value of the excess stress at the crack front. For example, we can suppose a proportionality between Σ_{11} and $\bar{\sigma}_{11}$,

$$\Sigma_{11} = D\bar{\sigma}_{11}. \tag{10.30}$$

In this case we have the following expression for the velocity of the crack front

$$V_C^2 = \frac{\bar{\sigma}_{11}(1+D)}{\rho_0 \left(1 + \bar{\sigma}_{11}/B\right)}. \tag{10.31}$$

Consequently,

$$\lim_{\sigma \to \infty} V_C^2 = \frac{B}{\rho_0} \left(1 + \lim_{\sigma \to \infty} \frac{\Sigma_{11}}{\bar{\sigma}_{11}}\right) = c_R^2 \left(1 + D\right). \tag{10.32}$$

It follows that in this case the limiting value of the velocity of the crack front is different from the value of the Rayleigh wave velocity. Denoting the limiting value of the velocity of the crack front by V_T, we can represent the expression for the velocity of the crack front as follows

$$V_C^2 = V_T^2 \left(1 + \frac{\rho_0 c_R^2 l}{AJ}\right)^{-1} = V_T^2 \left(1 - \left(1 + \frac{AJ}{\rho_0 c_R^2 l}\right)^{-1}\right). \tag{10.33}$$

To be able to compare the obtained relation with experimental data, we note that the value of the J-integral is proportional to the square of the stress intensity factor K_I in the considered problems [Fineberg and Marder (1999); Ravi-Chandar (2004)]. Therefore, we can rewrite the last expression in terms of the stress intensity factor

$$V_C^2 = V_T^2 \left(1 + \frac{Ml}{K_I^2}\right)^{-1} = V_T^2 \left(1 - \left(1 + \frac{K_I^2}{Ml}\right)^{-1}\right), \qquad (10.34)$$

where the coefficient M depends on the properties of the material.

Thus, the derived kinetic relation contains two model parameters: the limiting velocity V_T, which directly corresponds to the condition taken for the excess stress at the crack front, and the characteristic length scale l. Both model parameters may by adjusted to fit experimental data.

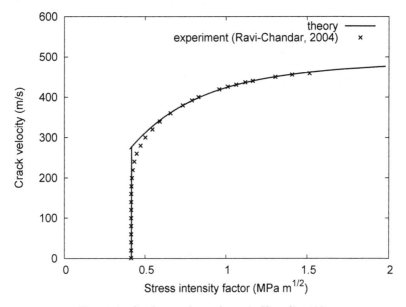

Fig. 10.4   Crack growth toughness in Homalite-100.

First, we compare the prediction (10.34) with the 'averaged' data for crack propagation in Homalite-100 given by Ravi-Chandar (2004). The result of the comparison is shown in Fig. 10.4, where the theoretical curve is calculated using the values $V_T = 500$ m/s and $Ml = 0.392$ MPa^{2}m. It should be noted that the expression (10.34) is applied only for the values $K_I > K_{Ic}$, and the critical value is $K_{Ic} = 0.415$ MPa m$^{1/2}$ (vertical line in the figure).

The 'averaged' $K_I - V_C$ relationship suggested by Ravi-Chandar (2004) is based on the experimental observations [Ravi-Chandar and Knauss (1984c)], where the crack velocity remained constant in each individual experiment. It is easy to see that for sufficiently small values of ML in Eq. (10.34), we will have almost constant crack velocity. Its limiting value

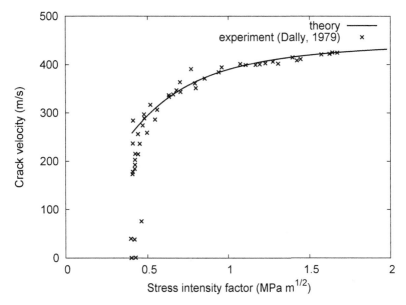

Fig. 10.5   Crack growth toughness in Homalite-100.

V_T appears to be dependent on the conditions of experiment.

Another comparison is made using experimental data for crack propagation in Homalite-100 by [Dally (1979)]. Here the value of the limiting velocity V_T is equal to 450 m/s and $Ml = 0.333$ MPa^{2}m. The result of the comparison is presented in Fig. 10.5, where the theoretical curve is shown again only for $K_I > K_{Ic}$. It seems that a good agreement between theory and experiment can be achieved by an appropriate choice of the values of the limiting velocity of the crack and of the characteristic length l.

One can suppose that the characteristic length l may be taken to be similar to the process zone length [Barenblatt (1996)]

$$l \sim \frac{K_{Ic}^2}{\sigma_*^2}. \tag{10.35}$$

In the thin strip geometry, it is possible to relate the values of σ_* and J [Hauch and Marder (1998)], which leads to

$$l \sim \frac{K_{Ic}^2}{J}. \tag{10.36}$$

In this case we arrive at an expression for the velocity of the crack front in the form

$$V_C^2 = V_T^2 \left(1 + \frac{M' K_{Ic}^2}{K_I^4}\right)^{-1} = V_T^2 \left(1 - \left(1 + \frac{K_I^4}{M' K_{Ic}^2}\right)^{-1}\right), \tag{10.37}$$

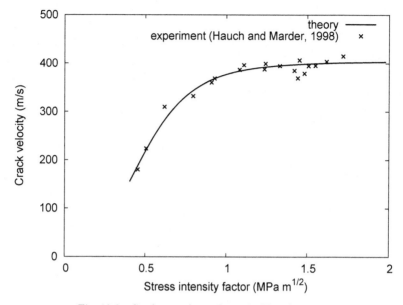

Fig. 10.6   Crack growth toughness in Homalite-100.

where M' is another material constant.

The last expression is useful to interpret the results of experiments for crack propagation in Homalite-100 by Hauch and Marder (1998), as one can see in Fig. 10.6. Here we have used the following values for the model parameters: $V_T = 405$ m/s and $M'K_{Ic}^2 = 0.154$ MPa^{4}m^2.

The same relation (10.37) was successfully applied for comparison of theoretical predictions and experimental data by Kobayashi and Mall (1978) (again for Homalite-100), which is shown in Fig. 10.7. Here another value of the limiting velocity $V_T = 385$ m/s was used but the value of the second model parameter was the same as for experiments by Hauch and Marder (1998).

As previously, a good fit of the experimental curves is obtained by the adjusting values of the limiting velocity and of the characteristic length.

10.5   Concluding remarks

We have applied a very similar formalism to the non-equilibrium description of the crack propagation as in the case of the phase-transition front dynamics. In a certain sense the formalism is more apparent in the case of

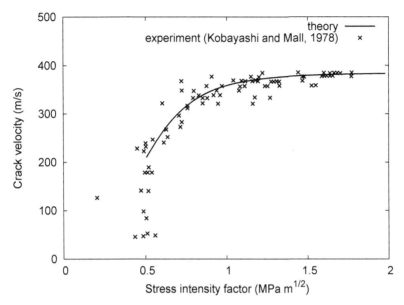

Fig. 10.7   Crack growth toughness in Homalite-100.

crack, because the values of jumps are reduced to the values of quantities characterizing a local equilibrium state in front of the crack front. The local equilibrium jump relation at the crack front allows us to establish a kinetic relation between the driving force and the velocity of a straight-through crack. This kinetic relation depends on the assumption concerning excess stress values at the crack front. It is shown that the obtained kinetic relation can be reduced to the prediction of the linear elasticity theory by setting the value of contact stress to zero. However, the adopted approach provides new possibilities. In particular, simple proportionality between excess and local equilibrium stresses leads to the limiting velocity, which is different from the Rayleigh wave speed. This assumption can be the subject of further modifications or generalizations.

The comparison of theoretical predictions with experimental data for Homalite-100 shows that in spite of the complexity of the problem and the high level of idealization of the model, the velocity of the crack can be estimated in the framework of the continuum theory. This can be confirmed once more by means of recent experimental data for the crack propagation in Polyester/TiO_2 nanocomposite by Evora, Jain and Shukla (2005), shown in Fig. 10.8. Here the theoretical curve is calculated using the measured value of the limiting velocity $V_T = 724$ m/s in Eq. (10.37). The second

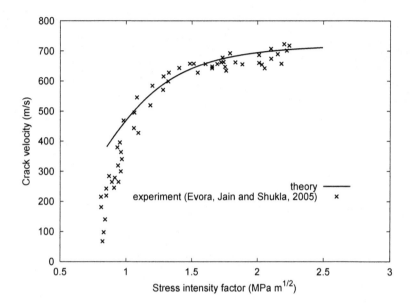

Fig. 10.8   Crack growth toughness in Polyester/TiO$_2$ nanocomposite.

model parameter corresponds to the critical value of the stress intensity factor $K_{Ic} = 0.85$ MPa m$^{1/2}$ [Evora, Jain and Shukla (2005)].

Chapter 11

Summing Up

Most dynamic processes at the macroscoppic level are governed by hyperbolic conservation laws. The theoretical description and numerical methods for the corresponding simulation of wave propagation in homogeneous materials are well developed and broadly applied. However, inhomogeneous materials are much more difficult to deal with although they can be formally modelled.

The real challenges met in the theoretical description and numerical simulation of propagation in inhomogeneous materials are well illustrated by the paradigmatic case of the propagation problem of transition fronts in the so-called martensitic alloys. The reason for this is that the dynamic response of a material capable of phase transformations is drastically different from that in the absence of phase change. Shape memory and pseudoelasticity of shape memory alloys are typical examples of the unusual behavior due to martensitic phase transformations. It is experimentally observed that martensitic phase transformation can be induced by an applied stress loading, e.g., a shock. In this case, phase-transition fronts may appear and move through the specimen. In the isothermal case, the governing equations for the motion are still hyperbolic conservation laws. However, these conservation laws are not able to predict the front appearance and velocity.

To understand the above set problem in depth, we have here applied the material description of continuum mechanics. This description allows us to determine the value of the driving force acting on the moving front and the full set of corresponding jump relations. However, the jump relations include the velocity of the front that remains undetermined. It happens that the velocity of the moving discontinuity is the most questionable property of the diffusionless martensitic phase-transition front.

Though it is clear that phase transformation processes are thermodynamically irreversible in general, the resulting entropy production cannot be calculated by means of classical phase equilibrium conditions. Therefore, the specially developed "local" equilibrium jump relations across the fronts are defined here and exploited satisfactorily. However, while the jump relations do not include the front velocity, they are formulated in terms of the so-called excess quantities. The excess quantities are needed to characterize the non-equilibrium states of the material. Fortunately, these excess quantities can be implemented in a finite volume numerical scheme as fluxes, which are determined at the boundaries between computational cells. The corresponding procedure is equivalent to the solution of a Riemann problem at the boundary between cells in the absence of moving fronts.

At discontinuities, we require that the jumps in the averaged fields keep the values of jumps in the genuine unknown fields. This is achieved by means of the continuity of excess quantities across the discontinuity. Adopting this assumption, we can establish a connection between the velocity of the front and the driving force acting on the front. This means that we derive the kinetic relation for the moving discontinuity.

Moreover, a criterion for the phase transformation initiation can also be formulated in the developed framework. This criterion is the condition of switching from one local equilibrium jump relation to another. The critical value of the driving force following from the initiation criterion allows us to predict the place and time of the initiation of the phase transformation.

Thus, to develop a numerical algorithm and to perform the numerical simulations of moving fronts we need to take into account:

- the material formulation of continuum mechanics, since we need to determine the driving force;
- a full thermomechanical description even in the isothermal case, because of the entropy production at the front;
- a non-equilibrium description of states of computational cells in the spirit of the thermodynamics of discrete systems;
- local equilibrium jump relations;
- a kinetic relation for the velocity of the phase boundary;
- the critical value of the driving force as the initiation criterion.

Throughout the book, we have discussed the formulated topics both from theoretical and numerical points of view, in one-dimensional and two-dimensional settings, in isothermal and adiabatic cases.

Similar analytical considerations are applied to the problem of the dynamics of straight brittle crack. The corresponding numerical simulations are supplemented by the needed calculation of the J-integral. The comparison with experimental data justifies the effectiveness and correctness of the applied framework for the description of wave and front propagation in inhomogeneous solids. This framework is based on the material formulation of continuum mechanics, the thermodynamics of discrete systems, and finite volume numerical methods. The theory is open for generalizations to include more complex material models, and the corresponding numerical algorithms should be improved in the case of complex geometry.

Appendix A

Thermodynamic interaction between two discrete systems in non-equilibrium[0]

There are two different descriptions of thermodynamic systems: the field formulation and the description as a discrete (or lumped) system. The field formulation or continuum thermodynamics deals with balance equations [Wilmanski (1988); Jou, Casas-Vazquez and Lebon (2001); Muschik, Papenfuss and Ehrentraut (2001)], which model together with the constitutive relations (equations of state) and the initial and boundary conditions of the process going on in the system.

After having inserted the constitutive relations into the balance equations, we obtain a system of partial differential equations whose analytical solutions can be calculated only in sufficiently simple cases. In numerous practical applications, these continuous models are replaced by approximations usually obtained by means of finite differences or finite elements. Calculations are carried out after having chosen an appropriate algorithm with respect to the problems of stability and convergence. Consequently for practical reasons it seems to be more convenient to have a direct description of the coupled thermodynamic behavior of a finite number of interacting elements or cells. This is just one of the reasons for introducing the concept of discrete systems [Muschik (1990)] because this gives a simple and effective possibility for describing thermodynamics and interactions of elements being in non-equilibrium. But here we encounter the analogous difficulties, which appears in classical thermodynamics, as noted by Truesdell and Bharatha [Truesdell and Bharatha (1977)] "the formal structure of classical thermodynamics describes the effects of changes undergone by

---

[0]This Appendix represents results of the paper by W. Muschik and A. Berezovski (Thermodynamic interaction between two discrete systems in non-equilibrium. *J. Non-Equilib. Thermodyn.*, **29**, (2004) 237–255) and is reprinted here by courtesy of Professor W. Muschik.

some single body. While it allows these effects for one body to be compared with corresponding effects for another body, it does not represent the effects associated with two bodies simultaneously or in any way conjointly." The interactions between a discrete system and its equilibrium environment are well known [Muschik (1993)], but the general interaction between two discrete non-equilibrium systems has not been investigated up to now and is shortly discussed here.

To describe the interaction between two discrete systems, we note that interacting systems always form a composite system. It is clear that the thermodynamic description of a composite system should be consistent with that of its interacting subsystems. It is also clear that the accuracy of the description depends on the information one has of the discrete system: If a compound system is described as being non-composed, the accuracy of the description is lower than taking its composition into account. The thermodynamic consistency of these two different descriptions is achieved by introducing excess quantities which vanish, if the discrete system is non-composed.

A.1   Equilibrium/non-equilibrium contacts

A.1.1   *Contact quantities*

By definition a *discrete (or lumped) system* is a region $G \in \mathbf{R}^3$ in space separated by a contact surface $\mathcal{F}$ from its environment G^* [Muschik (1993)]. The interaction between G and G^* can be described by exchange quantities. We call G a *Schottky system* [Schottky (1929)], if the interaction with its environment consists of heat exchange $\dot{Q}$, of power exchange $\dot{W}$, and of mass exchange described by time rates of mole numbers $\dot{\mathbf{n}}^e$ of the different species (Fig. A.1).

The exchange quantities $\dot{Q}$, $\dot{W}$, and $\dot{\mathbf{n}}^e$ determine intensive non-equilibrium contact quantities, namely the *contact temperature* Θ, the *dynamic pressure p* [1], and the *dynamic chemical potentials* $\boldsymbol{\mu}$ by different defining inequalities [Muschik (1993)]. These inequalities and their corresponding constraints are:

$$\dot{Q}(\Theta, T^*) \left(\frac{1}{\Theta} - \frac{1}{T^*} \right) \equiv \dot{Q} \left(\frac{1}{\Theta} - \frac{1}{T^*} \right) \geq 0, \quad (\dot{V} = 0, \ \dot{\mathbf{n}}^e = \mathbf{0}), \quad (A.1)$$

$$\dot{V}(p, p^*) \, (p - p^*) \equiv \dot{V} \, (p - p^*) \geq 0, \quad (\dot{Q} = 0, \ \dot{\mathbf{n}}^e = \mathbf{0}), \quad (A.2)$$

[1] as a special simple example for demonstration

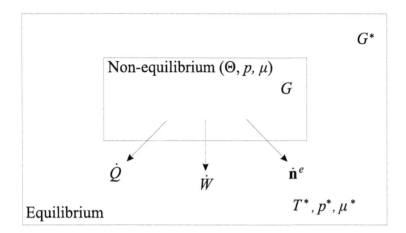

Fig. A.1 General structure of Schottky systems.

$$\dot{\mathbf{n}}^e(\boldsymbol{\mu}, \boldsymbol{\mu}^*) \cdot (\boldsymbol{\mu}^* - \boldsymbol{\mu}) \equiv \dot{\mathbf{n}}^e \cdot (\boldsymbol{\mu}^* - \boldsymbol{\mu}) \geq 0, \quad (\dot{Q} = 0, \ \dot{V} = 0). \qquad \text{(A.3)}$$

Here V is the volume of the discrete system G, T^* is the thermostatic temperature of the equilibrium environment G^*, p^* its equilibrium pressure, and $\boldsymbol{\mu}^*$ its equilibrium chemical potentials.

The contact temperature is the dynamical analogue to the thermostatic temperature [Muschik (1977); Muschik and Brunk (1977)]. The interpretation of the contact temperature is as follows: From Eq. (A.1) it is evident, that the heat exchange $\dot{Q}$ and the bracket always have the same sign [Muschik (1984)]. We now presuppose, that there exists exactly one equilibrium environment for each arbitrary state of a discrete system for which the net heat exchange between them vanishes. Consequently the defining inequality (A.1) determines the contact temperature Θ of the system as that thermostatic temperature T^* of the system's environment, for that this net heat exchange vanishes and no power and material exchanges occur. The same interpretation holds true for the dynamical pressure p and the dynamical chemical potentials $\boldsymbol{\mu}$, with respect to the net rate of the volume and the net rate of each mole number due to external material exchange. According to the *defining inequalities* (A.1) to (A.3) an equilibrium environment exists, for which the net heat exchange, the net power exchange and the net external material exchange for each component between G and G^* vanish. In this case the intensive variables $T^*, p^*, \boldsymbol{\mu}^*$ of this environment are defining the values of the non-equilibrium contact quantities $\Theta, p, \boldsymbol{\mu}$ of the system in consideration.

In all the cases the constraints mentioned in the brackets in Eqs. (A.1) to (A.3) have to be taken into consideration for each defining inequality. These defining inequalities are operational definitions of the contact quantities based on measuring rules which are expressed by the constraints mentioned in Eqs. (A.1) - (A.3). The contact quantities Θ, p, and $\boldsymbol{\mu}$ themselves are of course independent of these constraints and of the special values of T^*, p^*, and $\boldsymbol{\mu}^*$. Only for measuring them, the constraints have to be taken into consideration. Consequently Θ, p, and $\boldsymbol{\mu}$ are generally defined for Schottky systems.

Contact quantities satisfy *constitutive equations*

$$\dot{Q} = F\left(\frac{1}{\Theta} - \frac{1}{T^*}\right), \quad (\dot{V} = 0, \ \dot{\mathbf{n}}^e = \mathbf{0}), \tag{A.4}$$

$$\dot{V} = G\left(p - p^*\right), \quad (\dot{Q} = 0, \ \dot{\mathbf{n}}^e = \mathbf{0}), \tag{A.5}$$

$$\dot{\mathbf{n}}^e = \mathbf{H}\left(\boldsymbol{\mu}^* - \boldsymbol{\mu}\right), \quad (\dot{Q} = 0, \ \dot{V} = 0). \tag{A.6}$$

According to Eqs. (A.1) - (A.3) and their presupposed continuity the constitutive functions F, G, and $\mathbf{H}$ have the property

$$yF(y) \geq 0, \quad F(0) = 0, \tag{A.7}$$

$$yG(y) \geq 0, \quad G(0) = 0, \tag{A.8}$$

$$\mathbf{y} \cdot \mathbf{H}(\mathbf{y}) \geq 0, \quad \mathbf{H}(\mathbf{0}) = \mathbf{0}. \tag{A.9}$$

Beyond these properties, we additionally demand the strict monotony of F, G, and $\mathbf{H}$

$$y > y^+ \Rightarrow F(y) > F(y^+) \ \text{and} \ G(y) > G(y^+), \tag{A.10}$$

$$\mathbf{y} > \mathbf{y}^+ \Rightarrow \mathbf{H}(\mathbf{y}) > \mathbf{H}(\mathbf{y}^+). \tag{A.11}$$

The *non-equilibrium state space* of a discrete system in a rest frame can be chosen as follows [Muschik, Papenfuss and Ehrentraut (1997)]

$$Z = \{V, \mathbf{n}, U, \Theta, \boldsymbol{\xi} \ ; \ T^*, p^*, \boldsymbol{\mu}^*\}. \tag{A.12}$$

Here U is the internal energy of the system, and $\boldsymbol{\xi}$ are other variables in connection to irreversible processes going on in the system. Up to now it is not necessary to specify these variables. The dynamic pressure and the dynamical chemical potentials, and according to Eqs. (A.4) and (A.5) G and $\mathbf{H}$, are constitutive equations on Z (see sect. A.3.2, Eq. (A.65)). Thus, the complete description of the non-equilibrium state of a discrete system includes its contact quantities and the intensive variables of the environment as parameters which satisfy a Gibbs-Duhem equation, which means one of the $\boldsymbol{\mu}^*$ can be expressed by the other ones and by T^* and p^*.

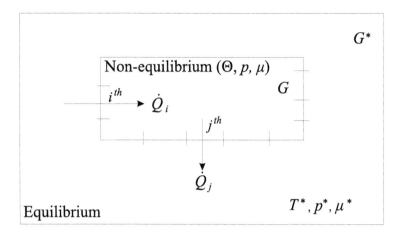

Fig. A.2   Division of the contact surface.

A.1.2   *Partial contact quantities*

The contact surface $\mathcal{F}$ between the system and its environment can be arbitrarily divided into parts $\mathcal{F}_i^+$ and $\mathcal{F}_j^-$ (Fig. A.2) defined by the following properties [Muschik (1984); Callen (1960)]:

$$\mathcal{F}^+ = \bigcup_i \mathcal{F}_i^+, \qquad \mathcal{F}_i^+ \cap \mathcal{F}_k^+ = \emptyset, \quad i \neq k, \tag{A.13}$$

$$\mathcal{F}^- = \bigcup_j \mathcal{F}_j^-, \qquad \mathcal{F}_j^- \cap \mathcal{F}_k^- = \emptyset, \quad j \neq k, \tag{A.14}$$

$$\mathcal{F} = \mathcal{F}^+ \cup \mathcal{F}^-, \qquad \mathcal{F}^+ \cap \mathcal{F}^- = \emptyset. \tag{A.15}$$

The denotion of the partial surfaces is determined by the signs of heat exchanges: $\dot{Q}_i^\pm$ are the heat exchanges through $\mathcal{F}_i^\pm$

$$\mathcal{F}_i^+: \quad \dot{Q}_i^+ \geq 0, \qquad \dot{Q}^+ := \sum_i \dot{Q}_i^+ \geq 0, \tag{A.16}$$

$$\mathcal{F}_j^-: \quad \dot{Q}_j^- < 0, \qquad \dot{Q}^- := \sum_j \dot{Q}_j^- < 0, \qquad \dot{Q}^+ + \dot{Q}^- = \dot{Q}. \tag{A.17}$$

We now consider a special state of G which we call *partial equilibrium with respect to p and $\boldsymbol{\mu}$*. This partial equilibrium of G is defined by

$$\mathcal{F}_k^\pm: \quad p_k = p^*, \qquad \boldsymbol{\mu} = \boldsymbol{\mu}^*, \quad \text{for all } k. \tag{A.18}$$

The partial equilibrium with respect to p and $\boldsymbol{\mu}$ means that there are no power and material exchanges through all the partial surfaces $\mathcal{F}_k^{\pm}$ according to Eqs. (A.2) and (A.3). Only heat exchanges occur. Consequently the heat exchanges satisfy the inequalities which belong to the partial contact surfaces according to Eq. (A.1)

$$\mathcal{F}_i^+: \qquad \dot{Q}_i^+ \left(\frac{1}{\Theta_i^+} - \frac{1}{T^*} \right) \geq 0, \qquad (p_i = p^*, \ \boldsymbol{\mu}_i = \boldsymbol{\mu}^*), \qquad \text{(A.19)}$$

$$\mathcal{F}_j^-: \qquad \dot{Q}_j^- \left(\frac{1}{\Theta_j^-} - \frac{1}{T^*} \right) \geq 0, \qquad (p_j = p^*, \ \boldsymbol{\mu}_j = \boldsymbol{\mu}^*). \qquad \text{(A.20)}$$

Summing up these inequalities, taking into consideration Eqs. (A.16) and (A.17), and applying the mean value theorem to the sums we obtain

$$\mathcal{F}^+: \qquad \sum_i \frac{\dot{Q}_i^+}{\Theta_i^+} - \frac{\dot{Q}^+}{T^*} \geq 0 \qquad \longrightarrow \qquad \dot{Q}^+ \left(\frac{1}{\Theta^+} - \frac{1}{T^*} \right) \geq 0, \qquad \text{(A.21)}$$

$$\mathcal{F}^-: \qquad \sum_j \frac{\dot{Q}_j^-}{\Theta_j^-} - \frac{\dot{Q}^-}{T^*} \geq 0 \qquad \longrightarrow \qquad \dot{Q}^- \left(\frac{1}{\Theta^-} - \frac{1}{T^*} \right) \geq 0. \qquad \text{(A.22)}$$

These inequalities are valid for arbitrary T^*, especially also for

$$T^* = \Theta, \qquad \longrightarrow \qquad \dot{Q}^+ + \dot{Q}^- = 0 \qquad \text{(A.23)}$$

which follows from Eq. (A.1). Hence Eqs. (A.21) and (A.22) result in

$$\mathcal{F}^+: \qquad \dot{Q}^+ \left(\frac{1}{\Theta^+} - \frac{1}{\Theta} \right) \geq 0 \qquad \longrightarrow \qquad \Theta^+ \leq \Theta, \qquad \text{(A.24)}$$

$$\mathcal{F}^-: \qquad \dot{Q}^+ \left(\frac{1}{\Theta} - \frac{1}{\Theta^-} \right) \geq 0 \qquad \longrightarrow \qquad \Theta \leq \Theta^-. \qquad \text{(A.25)}$$

Therefore we obtain for the partial equilibrium with respect to p and $\boldsymbol{\mu}$

$$\Theta^+ \leq \Theta \leq \Theta^-, \qquad (p_i = p^*, \ \boldsymbol{\mu}_i = \boldsymbol{\mu}^*, \text{ for all } i). \qquad \text{(A.26)}$$

Because Θ, Θ^+, and Θ^- are independent of the intensive variables of G^* and of the p_i and $\boldsymbol{\mu}^*$, the inequalities (A.26) are valid in general. This means the contact temperature Θ of $\mathcal{F}$ is always in between the contact temperatures Θ^+ and Θ^- of the partial contact surfaces $\mathcal{F}^+$ and $\mathcal{F}^-$.

From Eqs. (A.24) and (A.25) we obtain by Eqs. (A.16) and (A.17)

$$\frac{\dot{Q}^+}{\Theta^+} \geq \frac{\dot{Q}^+}{\Theta}, \qquad \frac{\dot{Q}^-}{\Theta^-} \geq \frac{\dot{Q}^-}{\Theta}. \qquad \text{(A.27)}$$

Summing up both inequalities we obtain by the use of Eqs. (A.21), (A.22), and (A.1)

$$\sum_k \frac{\dot{Q}_k}{\Theta_k} \geq \frac{\dot{Q}}{\Theta} \geq \frac{\dot{Q}}{T^*}. \tag{A.28}$$

Starting out with a partial equilibrium with respect to Θ and μ, we obtain from Eq. (A.2) by the same reasoning as before the generally valid inequalities analogously to Eq. (A.28)

$$\sum_k p_k \dot{V}_k \geq p\dot{V} \geq p^*\dot{V}. \tag{A.29}$$

Starting finally out with a partial equilibrium with respect to Θ and p we obtain from Eq. (A.3)

$$\sum_k \boldsymbol{\mu}_k \cdot \dot{\boldsymbol{n}}_k^e \leq \boldsymbol{\mu} \cdot \dot{\boldsymbol{n}}^e \leq \boldsymbol{\mu}^* \cdot \dot{\boldsymbol{n}}^e. \tag{A.30}$$

As for Eq. (A.4) the partial heat exchanges $\dot{Q}^+$ and $\dot{Q}^-$ satisfy constitutive equations which depend on the special partial contact surfaces $\mathcal{F}^\pm$ and on T^*

$$\dot{Q}^\pm = F_\pm \left(\frac{1}{\Theta^\pm} - \frac{1}{T^*} \right), \tag{A.31}$$

$$F_+(x) = 0, \quad \text{if } x \leq 0, \quad F_-(x) = 0, \quad \text{if } x \geq 0, \tag{A.32}$$

$$xF_\pm(x) \geq 0, \quad F_\pm(x) \text{ is strictly monotone, if not zero.} \tag{A.33}$$

According to Eqs. (A.5) and (A.6) there are constitutive equations of the rates of partial volumina and mole numbers analogous to Eq. (A.31).

The main consequence of this section is that partial contact quantities could be associated with the corresponding partial parts of the contact surface. The defined contact quantities provide the complete thermodynamic description of non-equilibrium states of discrete systems in an equilibrium environment [Muschik, Papenfuss and Ehrentraut (1997)]. However, in case of interacting non-equilibrium systems, we need some more concepts.

A.2 Interacting non-equilibrium systems

A.2.1 *Replacement quantities*

We now consider interacting non-equilibrium systems, a situation which differs from the previous one by a non-equilibrium environment instead of an equilibrium one. The first idea one may have is to replace the thermostatic temperature of the equilibrium environment by the contact temperature of the non-equilibrium environment. In this case it can be shown that the sign of the difference of the contact temperatures of the system and its environment does not determine the sign of the heat exchange [Muschik (1984)]. This means the heat exchange between these two non-equilibrium systems may not vanish, if the contact temperatures of both systems are equal. Therefore we replace the non-equilibrium environment by that equilibrium one which causes the same net heat exchange as in the original situation. This replacement is possible due to the monotony properties (A.10) and (A.11). We call the thermostatic temperature of that equilibrium environment the replacement temperature of the system's non-equilibrium environment in consideration.

Because the replacement introduces equilibrium environments instead of the non-equilibrium ones, the defining inequalities (A.1) - (A.3) are also valid for the replacement quantities: *replacement temperature* ϑ^*, *replacement pressure* π^*, and *replacement chemical potentials* $\boldsymbol{\nu}^*$ of the system's non-equilibrium environment,

$$\dot{Q}\left(\frac{1}{\Theta} - \frac{1}{\vartheta^*}\right) \geq 0, \quad (\dot{V} = 0, \ \dot{\mathbf{n}}^e = \mathbf{0}), \tag{A.34}$$

$$\dot{V}(p - \pi^*) \geq 0, \quad (\dot{Q} = 0, \ \dot{\mathbf{n}}^e = \mathbf{0}), \tag{A.35}$$

$$\dot{\mathbf{n}}^e \cdot (\boldsymbol{\nu}^* - \boldsymbol{\mu}) \geq 0, \quad (\dot{Q} = 0, \ \dot{V} = 0). \tag{A.36}$$

Here by definition the exchange quantities $\dot{Q}$, $\dot{V}$, and $\dot{\mathbf{n}}^e$ are the same as in the non-equilibrium situation in which two non-equilibrium systems are in contact with each other.

The difference between the inequalities (A.1) and (A.34) is as follows: In Eq. (A.1) T^* and $\dot{Q}$ are given and Θ is determined by the zero of $\dot{Q}$, whereas in Eq. (A.34) $\dot{Q}$ and Θ are given and ϑ^* is determined by them. The same is true for the other two inequalities (A.2) and (A.3), (A.35) and (A.36). Up to this different interpretation the inequalities (A.1) - (A.3) and (A.35) - (A.36) are formally identical.

Because of the equilibrium concept of replacement quantities, the constitutive functions F, G, and $\mathbf{H}$ in Eqs. (A.4) -(A.6) are also valid for the replacement quantities:

$$\dot{Q} = F\left(\frac{1}{\Theta} - \frac{1}{\vartheta^*}\right), \quad (\dot{V} = 0, \ \dot{\mathbf{n}}^e = \mathbf{0}), \tag{A.37}$$

$$\dot{V} = G\,(p - \pi^*)\,, \quad (\dot{Q} = 0, \ \dot{\mathbf{n}}^e = \mathbf{0}), \tag{A.38}$$

$$\dot{\mathbf{n}}^e = \mathbf{H}\,(\boldsymbol{\nu}^* - \boldsymbol{\mu})\,, \quad (\dot{Q} = 0, \ \dot{V} = 0). \tag{A.39}$$

The state space of a discrete system in a non-equilibrium environment should be changed, with respect to Eq. (A.12)

$$Z = \{V, \mathbf{n}, U, \Theta, \boldsymbol{\xi}\ ;\ \vartheta^*, \pi^*, \boldsymbol{\nu}^*\}. \tag{A.40}$$

Therefore, in the framework of thermodynamics of discrete systems, we can define all thermodynamic quantities necessary for the complete description of a discrete non-equilibrium system in a non-equilibrium environment.

A.2.2 *Composite systems*

Two discrete systems 1 and 2 interacting with each other are forming a *composite system* 1∪2, sometimes also called *compound system* (Fig. A.3). It is clear that the thermodynamic description of the composite system contains less information than that of the two subsystems themselves forming the compound system. We denote this fact as *compound deficiency*. This means that quantities belonging to the composite system differ from those belonging to the sum of both subsystems forming the composite system. The difference between these quantities is denoted as an *excess quantity*. In section A.3 the concepts of compound deficiency and excess quantities are worked out in more detail.

A.2.2.1 *The subsystems*

Suppose the considered composite system 1∪2 is composed of two subsystems 1 and 2 which are in interaction with each other and with the environment (marked by *) which is the same for both (Fig. A.3). As usual for Schottky systems the interaction consists of heat-, power-, and mass-exchange. Here especially the power-exchange is chosen as a volume work for simplification. In general the subsystems are in non-equilibrium whereas the environment is presupposed to be in equilibrium because of its reservoir properties.

For describing the exchanges we have to introduce three *contact surfaces*, one between both the subsystems 1 and 2, called $\mathcal{F}$ and two other ones between one of the two subsystems and the environment, denoted by $\mathcal{F}_1$ and $\mathcal{F}_2$ (see Fig. A.3). For the three contact surfaces we introduce four *partial contact temperatures* belonging to the subsystems 1 and 2 [Callen

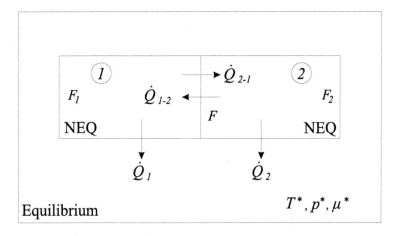

Fig. A.3   Systems 1 and 2 as parts of a composite system 1∪2.

(1960)] $\Theta_1^{\mathcal{F}}$, $\Theta_2^{\mathcal{F}}$, $\Theta^{\mathcal{F}_1}$, and $\Theta^{\mathcal{F}_2}$. The meaning of these contact temperatures is clear: $\Theta_j^{\mathcal{K}}$ is the partial contact temperature belonging to the contact surface $\mathcal{K}$ and j marks the subsystem, if necessary. The defining inequalities for these contact temperatures are

$$\dot{Q}_1 \left(\frac{1}{\Theta^{\mathcal{F}_1}} - \frac{1}{T^*} \right) \geq 0, \tag{A.41}$$

$$\dot{Q}_2 \left(\frac{1}{\Theta^{\mathcal{F}_2}} - \frac{1}{T^*} \right) \geq 0, \tag{A.42}$$

whereas we obtain according to Eq. (A.34) the following inequalities for the internal contact surface $\mathcal{F}$

$$\dot{Q}_{1-2} \left(\frac{1}{\Theta_1^{\mathcal{F}}} - \frac{1}{\vartheta_2} \right) \geq 0, \tag{A.43}$$

$$\dot{Q}_{2-1} \left(\frac{1}{\Theta_2^{\mathcal{F}}} - \frac{1}{\vartheta_1} \right) \geq 0. \tag{A.44}$$

According to Eq. (A.34) ϑ_1 and ϑ_2 are the replacement temperatures belonging to the non-equilibrium subsystems 1 and 2.

Because the contact surface between both subsystems is an *inert* one, which means that heat and mass are not absorbed or emitted by this partition[2], we have ($\mathbf{h}_j$ are the molar enthalpies of the j-th subsystem)

---

[2]more details for open systems see [Muschik and Gümbel (1999)]

$$\dot{Q}_{1-2} + \mathbf{h}_1 \cdot \dot{\mathbf{n}}_{1-2}^e = -\dot{Q}_{2-1} - \mathbf{h}_2 \cdot \dot{\mathbf{n}}_{2-1}^e, \qquad (A.45)$$

and

$$\dot{\mathbf{n}}_{1-2}^e = -\dot{\mathbf{n}}_{2-1}^e. \qquad (A.46)$$

Hence Eq. (A.44) results in

$$[-\dot{Q}_{1-2} - (\mathbf{h}_1 - \mathbf{h}_2) \cdot \dot{\mathbf{n}}_{1-2}^e] \left(\frac{1}{\Theta_2^{\mathcal{F}}} - \frac{1}{\vartheta_1} \right) \geq 0. \qquad (A.47)$$

Because this inequality is valid for arbitrary $\dot{\mathbf{n}}_{1-2}^e$, especially also for $\dot{\mathbf{n}}_{1-2}^e = \mathbf{0}$, and $\dot{Q}_{1-2}, \Theta_2^{\mathcal{F}}$, and ϑ_1 are independent of $\dot{\mathbf{n}}_{1-2}^e$, we obtain from Eqs. (A.43) and (A.47)

$$\operatorname{sign} \left(\frac{1}{\Theta_1^{\mathcal{F}}} - \frac{1}{\vartheta_2} \right) = -\operatorname{sign} \left(\frac{1}{\Theta_2^{\mathcal{F}}} - \frac{1}{\vartheta_1} \right). \qquad (A.48)$$

The *entropy production* generated by the heat exchange between the subsystems of the compound system $1 \cup 2$ is the left-hand side of Eq. (A.43), or equivalently that of Eq. (A.44), because the contact surface $\mathcal{F}$ between them is inert. Therefore the left-hand sides of Eqs. (A.43) and (A.44) are equal and Eq. (A.48) results in

$$\frac{1}{\Theta_1^{\mathcal{F}}} - \frac{1}{\vartheta_2} = \frac{1}{\vartheta_1} - \frac{1}{\Theta_2^{\mathcal{F}}}. \qquad (A.49)$$

By this equation the mean value of the reciprocal replacement temperatures is determined by that of the reciprocal contact temperatures

$$\frac{1}{\tau^{\mathcal{F}}} := \frac{1}{2} \left(\frac{1}{\Theta_1^{\mathcal{F}}} + \frac{1}{\Theta_2^{\mathcal{F}}} \right) = \frac{1}{2} \left(\frac{1}{\vartheta_1} + \frac{1}{\vartheta_2} \right). \qquad (A.50)$$

This relation shows the replacement temperatures of two non-equilibrium systems being in thermal contact with each other are dependent of each other in contrast to the independent contact temperatures. If one of the contacting subsystems (say 2) is in equilibrium with the environment, we obtain from Eq. (A.50)

$$\Theta_2^{\mathcal{F}} = \vartheta_2 = T^* \longrightarrow \vartheta_1 = \Theta_1^{\mathcal{F}}. \qquad (A.51)$$

A.2.2.2 *The composite system*

In this section we describe the composite system $1 \cup 2$, as if it would not be composed of the two subsystems 1 and 2. This description is of course a more coarse one than that of the two single subsystems in interaction.

As already mentioned the subsystems of the composite system have two contact surfaces with respect to the environment, $\mathcal{F}_1$ and $\mathcal{F}_2$, to which the two contact temperatures $\Theta^{\mathcal{F}_1}$ and $\Theta^{\mathcal{F}_2}$ belong (see Fig. 3). The contact surface of the composite system is $\mathcal{F}_1 \cup \mathcal{F}_2$ to which the contact temperature Θ belongs. Without restricting generality we presuppose that

$$\Theta^{\mathcal{F}_1} \leq \Theta^{\mathcal{F}_2} \tag{A.52}$$

is valid, because this inequality only depends on the numbering of the subsystems. In this case the inequalities (A.26) and (A.28) become

$$\Theta^{\mathcal{F}_1} \leq \Theta \leq \Theta^{\mathcal{F}_2}, \tag{A.53}$$

$$\sum_{k=1}^{2} \frac{\dot{Q}^{\mathcal{F}_k}}{\Theta^{\mathcal{F}_k}} \geq \frac{\dot{Q}}{\Theta} \geq \frac{\dot{Q}}{T^*}, \qquad \dot{Q} = \dot{Q}^{\mathcal{F}_1} + \dot{Q}^{\mathcal{F}_2}, \tag{A.54}$$

and we obtain from the inequalities (A.29) and (A.30)

$$\sum_{k=1}^{2} p^{\mathcal{F}_k} \dot{V}^{\mathcal{F}_k} \geq p\dot{V} \geq p^*\dot{V}, \qquad \dot{V} = \dot{V}^{\mathcal{F}_1} + \dot{V}^{\mathcal{F}_2}, \quad ^3 \tag{A.55}$$

$$\sum_{k=1}^{2} \boldsymbol{\mu}^{\mathcal{F}_k} \cdot \dot{\boldsymbol{n}}^{e^{\mathcal{F}_k}} \leq \boldsymbol{\mu} \cdot \dot{\boldsymbol{n}}^e \leq \boldsymbol{\mu}^* \cdot \dot{\boldsymbol{n}}^e, \qquad \dot{\boldsymbol{n}}^e = \dot{\boldsymbol{n}}^{e^{\mathcal{F}_1}} + \dot{\boldsymbol{n}}^{e^{\mathcal{F}_2}}. \tag{A.56}$$

Now we have defined both kinds of contact quantities, those for the composite system itself and those for its subsystems. The inequalities (A.54) and (A.56) characterize what we define as *compound deficiency*. This means a more detailed description by the subsystems yields other results than a coarse description by the composite system.

A.3   Compound deficiency

A.3.1   *The inequalities*

We now consider the compound deficiency of the different descriptions of the composite system and its subsystems (Fig. A.3).

The power exchange between the two subsystems and the environment is

$$\dot{W} := \dot{W}_1 + \dot{W}_2 = -\sum_{k=1}^{2} p^{\mathcal{F}_k} \dot{V}^{\mathcal{F}_k}, \tag{A.57}$$

---

^{3}the rates of the partial volumina $\dot{V}^{\mathcal{F}_k}$ can be defined properly by using Reynolds transport theorem whereas the power exchange between the composite systems and the environment is by taking Eq. (A.55) into account

$$\dot{W}_{CS} := -p\dot{V} \geq \dot{W}. \tag{A.58}$$

For the energy exchange due to mass exchange between the two subsystems and the environment follows from Eq. (A.56)

$$\dot{M} := \dot{M}_1 + \dot{M}_2 = \sum_{k=1}^{2} \boldsymbol{\mu}^{\mathcal{F}_k} \cdot \dot{\boldsymbol{n}}^{e\mathcal{F}_k}. \tag{A.59}$$

The corresponding energy exchange due to mass exchange between the compound system and the environment is

$$\dot{M}_{CS} := \boldsymbol{\mu} \cdot \dot{\boldsymbol{n}}^e \geq \dot{M}. \tag{A.60}$$

The last inequality follows from Eq. (A.56). In section A.3.3 we investigate how these compound deficiency inequalities transform to other thermodynamical quantities.

A.3.2 *Energy and entropy*

The first laws of the composite system and its environment are

$$\dot{U} = \dot{Q} - p\dot{V} + \mathbf{h} \cdot \dot{\mathbf{n}}^e, \qquad \dot{U}^* = \dot{Q}^* - p^*\dot{V}^* + \mathbf{h}^* \cdot \dot{\mathbf{n}}^{e*}. \tag{A.61}$$

The composite system is separated from its environment by an *inert partition* which is characterized by the following properties [see Eqs. (A.45), (A.46)]

$$\dot{Q} + \mathbf{h} \cdot \dot{\mathbf{n}}^e = -\dot{Q}^* - \mathbf{h}^* \cdot \dot{\mathbf{n}}^{e*}, \tag{A.62}$$

$$\dot{\mathbf{n}}^* = \dot{\mathbf{n}}^{e*} = -\dot{\mathbf{n}}^e, \qquad \dot{V} = -\dot{V}^*, \qquad (\dot{\mathbf{n}} = \dot{\mathbf{n}}^e + \dot{\mathbf{n}}^i, \ \dot{\mathbf{n}}^{i*} \equiv \mathbf{0}) \tag{A.63}$$

(^{e} denotes external exchange, whereas ^{i} marks the change by chemical reactions).

Therefore Eq. (A.61)$_2$ yields

$$\dot{U}^* = -\dot{Q} - p^*\dot{V}^* - \mathbf{h} \cdot \dot{\mathbf{n}}^e. \tag{A.64}$$

The reduced heat exchange $\dot{Q}/\Theta$ in Eq. (A.54) is connected to the entropy rate of a discrete non-equilibrium system [Muschik (1993)]

$$\dot{S} = \frac{1}{\Theta}\left(\dot{U} + p\dot{V} - \boldsymbol{\mu} \cdot \dot{\mathbf{n}}\right) + \alpha\dot{\Theta} + \boldsymbol{\beta} \cdot \dot{\boldsymbol{\xi}}. \tag{A.65}$$

Here $\dot{\boldsymbol{\xi}}$ are the time rates of the variables in the state space (Eq. (A.12)) characterizing irreversible processes. An example for such variables are the chemical reaction rates. In this case the $\boldsymbol{\beta}$ are the affinities of the chemical reactions.

The relation (A.65) represents a non-equilibrium extension of Gibbs fundamental equation which for the equilibrium environment is as follows

$$\dot{S}^* = \frac{1}{T^*} \left(\dot{U}^* + p^* \dot{V}^* - \boldsymbol{\mu}^* \cdot \dot{\mathbf{n}}^* \right). \tag{A.66}$$

The rates $\dot{\Theta}$ and $\dot{\boldsymbol{\xi}}$ describe non-equilibrium. The non-equilibrium entropy is a function $S(U, V, \mathbf{n}, \Theta, \boldsymbol{\xi})$ on the state space [Muschik (1993)]. The equilibrium entropy $S^*(U^*, V^*, \mathbf{n}^*)$ depends on the equilibrium variables of the environment.

Introducing Eqs. (A.61)$_1$ and (A.64) into Eqs. (A.65) and (A.66) we obtain

$$\dot{S} = \frac{1}{\Theta} \left(\dot{Q} + \mathbf{h} \cdot \dot{\mathbf{n}}^e - \boldsymbol{\mu} \cdot \dot{\mathbf{n}} \right) + \alpha \dot{\Theta} + \boldsymbol{\beta} \cdot \dot{\boldsymbol{\xi}}, \tag{A.67}$$

$$\dot{S}^* = \frac{1}{T^*} \left(-\dot{Q} - \mathbf{h} \cdot \dot{\mathbf{n}}^e + \boldsymbol{\mu}^* \cdot \dot{\mathbf{n}}^e \right). \tag{A.68}$$

By the molar entropies

$$\Theta \mathbf{s} := \mathbf{h} - \boldsymbol{\mu}, \qquad T^* \mathbf{s}^* := \mathbf{h}^* - \boldsymbol{\mu}^* \tag{A.69}$$

Eqs. (A.67) and (A.68) result in

$$\dot{S} = \frac{1}{\Theta} \left(\dot{Q} + \Theta \mathbf{s} \cdot \dot{\mathbf{n}}^e - \boldsymbol{\mu} \cdot \dot{\mathbf{n}}^i \right) - \alpha \dot{\Theta} + \boldsymbol{\beta} \cdot \dot{\boldsymbol{\xi}}, \tag{A.70}$$

$$\dot{S}^* = \frac{1}{T^*} \left(-\dot{Q} - \Theta \mathbf{s} \cdot \dot{\mathbf{n}}^e + (\boldsymbol{\mu}^* - \boldsymbol{\mu}) \cdot \dot{\mathbf{n}}^e \right). \tag{A.71}$$

Presupposing additivity of entropies we obtain for the entropy rate of the isolated total system

$$\dot{S}^{tot} = \dot{S} + \dot{S}^* = \left(\frac{1}{\Theta} - \frac{1}{T^*} \right) \left(\dot{Q} + \Theta \mathbf{s} \cdot \dot{\mathbf{n}}^e \right)$$
$$+ \frac{1}{T^*} (\boldsymbol{\mu}^* - \boldsymbol{\mu}) \cdot \dot{\mathbf{n}}^e - \frac{\boldsymbol{\mu}}{\Theta} \cdot \dot{\mathbf{n}}^i + \alpha \dot{\Theta} + \boldsymbol{\beta} \cdot \dot{\boldsymbol{\xi}} \geq 0. \tag{A.72}$$

Here the inequality is caused by the second law valid for isolated systems for which entropy rate and entropy production are identical.

If there is no heat and mass exchange between the system and its environment, Eq. (A.72) results in the entropy production of the system in consideration

$$-\frac{\mu}{\Theta} \cdot \dot{\mathbf{n}}^i + \alpha \dot{\Theta} + \beta \cdot \dot{\xi} \geq 0. \tag{A.73}$$

Since this entropy production is independent of the intensive variables of the equilibrium environment, from Eq. (A.72) the following inequality follows

$$\left(\frac{1}{\Theta} - \frac{1}{T^*}\right)\left(\dot{Q} + \Theta \mathbf{s} \cdot \dot{\mathbf{n}}^e\right) + \frac{1}{T^*}(\mu^* - \mu) \cdot \dot{\mathbf{n}}^e \geq 0. \tag{A.74}$$

If $\dot{\mathbf{n}}^e = \mathbf{0}$, we obtain Eq. (A.1), if $T^* = \Theta$ Eq. (A.3) follows. Thus the defining inequalities are rediscovered.

For discussing compound deficiency in more detail we consider a special example in the next section.

A.3.3 *Example: An endoreversible system*

In the sequel we consider an endoreversible system which by definition consists of subsystems being in different equilibria [Hoffmann, Burzler and Schubert (1997)]. Because these equilibria are different, irreversible processes take place between the subsystems. The use of endoreversible systems represents an analogue for discrete systems of the often accepted hypothesis of local equilibrium of classical irreversible thermodynamics [Keller (1971); Muschik (1979); Kestin (1992, 1993); Maugin and Muschik (1994); Maugin (1999b); De Groot and Mazur (1961)]. Using it, points of the non-equilibrium state space are associated with points of equilibrium subspace by means of a projection. Thus we suppose a non-equilibrium state of each discrete system is associated with an equilibrium state of the accompanying reversible process.

In an endoreversible system all contact quantities of the subsystems are identified with equilibrium bulk values

$$\Theta^{\mathcal{F}_j} = \Theta^{\mathcal{F}}_j = \vartheta_j =: T_j, \qquad j = 1, 2. \tag{A.75}$$

For local equilibrium Eq. (A.50) is satisfied trivially.

In case of an endoreversible system the state space is much smaller than that in non-equilibrium (Eq. (A.12)) [Muschik, Papenfuss and Ehrentraut (1997)]

$$Z^{eq} = \{V, \mathbf{n}, T\}. \tag{A.76}$$

For the two subsystems in consideration the local accompanying states are

$$Z_j^{eq} = \{V_j, \mathbf{n}_j, T_j\}, \qquad T_1 \neq T_2, \quad V_1 \neq V_2, \quad \mathbf{n}_1 \neq \mathbf{n}_2. \qquad (A.77)$$

According to this, the entropies of the equilibrium subsystems 1 and 2 are (Fig. A.3)

$$\dot{S}_1 = \frac{1}{T_1}\dot{Q}_1 + \mathbf{s}_1 \cdot \dot{\mathbf{n}}_1^e + \frac{\dot{Q}_{1-2}}{T_1} + \mathbf{s}_1 \cdot \dot{\mathbf{n}}_{1-2}^e, \qquad (A.78)$$

$$\dot{S}_2 = \frac{1}{T_2}\dot{Q}_2 + \mathbf{s}_2 \cdot \dot{\mathbf{n}}_2^e + \frac{\dot{Q}_{2-1}}{T_2} + \mathbf{s}_2 \cdot \dot{\mathbf{n}}_{2-1}^e. \qquad (A.79)$$

Because the partition between the two subsystems is an inert one, we obtain according to Eqs. (A.45) and (A.46) for the heat exchanges of both subsystems

$$\dot{Q}_{1-2} + \mathbf{h}_1 \cdot \dot{\mathbf{n}}_{1-2}^e = -\dot{Q}_{2-1} - \mathbf{h}_2 \cdot \dot{\mathbf{n}}_{2-1}^e \qquad \dot{\mathbf{n}}_{2-1}^e = -\dot{\mathbf{n}}_{1-2}^e. \qquad (A.80)$$

Inserting Eq. (A.80) into Eq. (A.79) yields

$$\dot{S} := \dot{S}_1 + \dot{S}_2 = \frac{\dot{Q}_1}{T_1} + \frac{\dot{Q}_2}{T_2} + \mathbf{s}_1 \cdot \dot{\mathbf{n}}_1^e + \mathbf{s}_2 \cdot \dot{\mathbf{n}}_2^e$$
$$+ \left(\frac{1}{T_1} - \frac{1}{T_2}\right)\dot{Q}_{1-2} + (\mathbf{s}_1 - \mathbf{s}_2) \cdot \dot{\mathbf{n}}_{1-2}^e + \frac{1}{T_2}(\mathbf{h}_2 - \mathbf{h}_1) \cdot \dot{\mathbf{n}}_{1-2}^e. \qquad (A.81)$$

The composite system $1 \cup 2$ is an endoreversible non-equilibrium system whose entropy rate is according to Eq. (A.70)

$$\dot{S}_{CS} := \frac{1}{\Theta}\dot{Q} + \mathbf{s} \cdot \dot{\mathbf{n}}^e. \qquad (A.82)$$

For the internal energy we obtain according to Eq. (A.61)

$$\dot{U}_1 = \dot{Q}_1 - p_1\dot{V}_1 + \mathbf{h}_1 \cdot \dot{\mathbf{n}}_1^e + \dot{Q}_{1-2} - p_1\dot{V}_{1-2} + \mathbf{h}_1 \cdot \dot{\mathbf{n}}_{1-2}^e, \qquad (A.83)$$

$$\dot{U}_2 = \dot{Q}_2 - p_2\dot{V}_2 + \mathbf{h}_2 \cdot \dot{\mathbf{n}}_2^e + \dot{Q}_{2-1} - p_2\dot{V}_{2-1} + \mathbf{h}_2 \cdot \dot{\mathbf{n}}_{2-1}^e. \qquad (A.84)$$

Presupposing the additivity of partial energies yields, if Eq. (A.80) is taken into account,

$$\dot{U} := \dot{U}_1 + \dot{U}_2 = \dot{Q}_1 + \dot{Q}_2 - p_1\dot{V}_1 - p_2\dot{V}_2$$
$$+ \mathbf{h}_1 \cdot \dot{\mathbf{n}}_1^e + \mathbf{h}_2 \cdot \dot{\mathbf{n}}_2^e - (p_1 - p_2)\dot{V}_{1-2}. \qquad (A.85)$$

For the endoreversible non-equilibrium composite system we obtain from Eq. (A.61)

$$\dot{U}_{CS} = \dot{Q} - p\dot{V} + \mathbf{h} \cdot \dot{\mathbf{n}}^e. \qquad (A.86)$$

A comparison of entropies (Eq. (A.82)) and energies (Eq. (A.86)) of the composite system with those of the subsystems (Eqs. (A.81) and (A.85)) shows that they are different

$$\dot{S}_{CS} \neq \dot{S}, \qquad \dot{U}_{CS} \neq \dot{U}. \tag{A.87}$$

This fact is just what we denoted by compound deficiency. Also the inequalities (A.58) and (A.60) are caused by compound deficiency. The differences between the quantities belonging to the composite system and those belonging to the subsystems are discussed in the next section.

A.3.4 *Excess quantities*

A.3.4.1 *Excess power exchange, excess mass exchange*

For describing compound deficiency in more detail we introduce *excess quantities*. If we denote a special quantity by $\square$, the corresponding excess quantity $\square^{EX}$ is defined by

$$\sum_k \square_k + \square^{EX} := \square_{CS}. \tag{A.88}$$

We now discuss some of these excess quantities.

From Eq. (A.55) we obtain the *excess power exchange*

$$\dot{W}^{EX} = -p\dot{V} + p_1\dot{V}_1 + p_2\dot{V}_2 \geq 0. \tag{A.89}$$

If the discrete system in consideration is non-composed, that means, if $p_1 = p_2 = p$ is valid, we obtain $\dot{W}^{EX} = 0$.

The *excess energy exchange* due to mass exchange is according to Eq. (A.56)

$$\dot{M}^{EX} = \boldsymbol{\mu} \cdot \dot{\mathbf{n}}^e - \boldsymbol{\mu}_1 \cdot \dot{\mathbf{n}}_1^e + \boldsymbol{\mu}_2 \cdot \dot{\mathbf{n}}_2^e \geq 0. \tag{A.90}$$

If the discrete system in consideration is non-composed, this means, if $\boldsymbol{\mu}_1 = \boldsymbol{\mu}_2 = \boldsymbol{\mu}$ is valid, we obtain $\dot{M}^{EX} = 0$.

A.3.4.2 *Excess energy*

According to Eq. (A.88) the rate of the *excess energy* is by using Eqs. (A.86) and (A.85)

$$\begin{aligned} \dot{U}^{EX} :=& \dot{U}_{CS} - \dot{U}_1 - \dot{U}_2 \;=\; \dot{Q} - p\dot{V} + \mathbf{h} \cdot \dot{\mathbf{n}}^e \\ & -\dot{Q}_1 - \dot{Q}_2 + p_1\dot{V}_1 + p_2\dot{V}_2 - \mathbf{h}_1 \cdot \dot{\mathbf{n}}_1^e - \mathbf{h}_2 \cdot \dot{\mathbf{n}}_2^e + (p_1 - p_2)\dot{V}_{1-2}. \end{aligned} \tag{A.91}$$

Taking Eqs. (A.17)$_3$ and (A.89) into account we obtain

$$\dot{U}^{EX} = \dot{W}^{EX} + (\mathbf{h} - \mathbf{h}_1) \cdot \dot{\mathbf{n}}_1^e + (\mathbf{h} - \mathbf{h}_2) \cdot \dot{\mathbf{n}}_2^e + (p_1 - p_2)\dot{V}_{1-2}. \quad (A.92)$$

If the discrete system in consideration is non-composed, that means, if $\mathbf{h}_1 = \mathbf{h}_2 = \mathbf{h}$ and $p_1 = p_2$ are valid, we obtain $\dot{U}^{EX} = 0$.

A.3.4.3   *Excess entropy*

According to Eq. (A.88) the rate of the *excess entropy* is by using Eqs. (A.82) and (A.81)

$$
\begin{aligned}
\dot{S}^{EX} := \dot{S}_{CS} - \dot{S}_1 - \dot{S}_2 \; = \; & \frac{1}{\Theta}\dot{Q} + \mathbf{s} \cdot \dot{\mathbf{n}}^e - \frac{\dot{Q}_1}{T_1} - \frac{\dot{Q}_2}{T_2} \\
& - \mathbf{s}_1 \cdot \dot{\mathbf{n}}_1^e - \mathbf{s}_2 \cdot \dot{\mathbf{n}}_2^e - \left(\frac{1}{T_1} - \frac{1}{T_2}\right)\dot{Q}_{1-2} \quad (A.93) \\
& - (\mathbf{s}_1 - \mathbf{s}_2) \cdot \dot{\mathbf{n}}_{1-2}^e - \frac{1}{T_2}(\mathbf{h}_2 - \mathbf{h}_1) \cdot \dot{\mathbf{n}}_{1-2}^e.
\end{aligned}
$$

Using Eq. (A.69) we obtain after a short calculation

$$
\begin{aligned}
\dot{S}^{EX} = \; & \left(\frac{1}{\Theta} - \frac{1}{T_1}\right)\dot{Q}_1 + \left(\frac{1}{\Theta} - \frac{1}{T_2}\right)\dot{Q}_2 \\
& + (\mathbf{s} - \mathbf{s}_1) \cdot \dot{\mathbf{n}}_1^e + (\mathbf{s} - \mathbf{s}_2) \cdot \dot{\mathbf{n}}_2^e \quad (A.94) \\
& - \left(\frac{1}{T_1} - \frac{1}{T_2}\right)\left(\dot{Q}_{1-2} + \mathbf{h}_1 \cdot \dot{\mathbf{n}}_{1-2}^e\right) - \left(\frac{\mu_2}{T_2} - \frac{\mu_1}{T_1}\right) \cdot \dot{\mathbf{n}}_{1-2}^e.
\end{aligned}
$$

If the discrete system in consideration is non-composed, that means, if $T_1 = T_2 = \Theta$, $\mathbf{s}_1 = \mathbf{s}_2 = \mathbf{s}$ and $\mu_1 = \mu_2$ are valid, we obtain $\dot{S}^{EX} = 0$.

A.4   Concluding remarks

A discrete system may be composed of subsystems interacting with each other or may be non-composed. Therefore the description of the discrete system in consideration depends on the information one has about the system. It is clear the two descriptions are different, because different levels of information result in different contact quantities between the considered discrete system and its equilibrium environment. So e.g. the contact temperatures depend on the description: If the discrete system is considered as being non-composed, the contact temperature is that of the whole contact surface between the discrete system and its environment. If the discrete system is described as a composite system, the contact temperatures of the partial contact surfaces between the subsystems of the composite system and the environment play a role.

The difference of these descriptions is a general feature in thermodynamics of discrete systems which we characterize by the concept of compound deficiency. Starting out with the additivity of partial exchange quantities the compound deficiency results in different net exchange quantities and different energies and entropies depending on the description of the considered discrete system as being composed or not. The thermodynamic consistency between the different descriptions is achieved by introducing excess quantities.

As it is shortly discussed, the concept of compound deficiency can be extended to the contact of two discrete non-equilibrium systems by introducing replacement quantities instead of contact quantities. The excess quantities of energy, entropy, and power and energy exchanges are calculated and discussed for an endoreversible compound system.

Bibliography

Abeyaratne, R., Bhattacharya, K. and Knowles, J. K. (2001). Strain-energy functions with local minima: Modeling phase transformations using finite thermoelasticity, in Y. Fu and R.W. Ogden (eds.), *Nonlinear Elasticity: Theory and Application* (Cambridge University Press), pp. 433–490.

Abeyaratne, R. and Knowles, J. K. (1990). On the driving traction acting on a surface of strain discontinuity in a continuum, *J. Mech. Phys. Solids* **38**, pp. 345–360.

Abeyaratne, R. and Knowles, J. K. (1991). Kinetic relations and the propagation of phase boundaries in solids, *Arch. Rat. Mech. Anal.* **114**, pp. 119–154.

Abeyaratne, R. and Knowles, J. K. (1993). A continuum model of a thermoelastic solid capable of undergoing phase transitions. *J. Mech. Phys. Solids* **41**, pp. 541–571.

Abeyaratne, R. and Knowles, J. K. (1994a). Dynamics of propagating phase boundaries: thermoelastic solids with heat conduction, *Arch. Rational Mech. Anal.* **126**, pp, 203–230.

Abeyaratne, R. and Knowles, J. K. (1994b). Dynamics of propagating phase boundaries: Adiabatic theory for thermoelastic solids, *Physica D* **79**, pp. 269–288.

Abeyaratne, R. and Knowles, J. K. (1997a). Impact-induced phase transitions in thermoelastic solids, *Phil. Trans. R. Soc. Lond.* **A355**, pp. 843–867.

Abeyaratne, R. and Knowles, J. K. (1997b). On the kinetics of an austenite-martensite phase transformation induced by impact in a Cu-Al-Ni shape-memory alloy, *Acta Mater.* **45**, pp. 1671–1683.

Abeyaratne, R. and Knowles, J. K. (2000). A note on the driving traction acting on a propagating interface: Adiabatic and non-adiabatic processes of a continuum, *ASME J. Appl. Mech.* **67**, pp. 829–831.

Abeyaratne, R. and Knowles, J. K. (2006). *Evolution of Phase Transitions : A Continuum Theory* (Cambridge University Press).

Achenbach, J. D. (1973). *Wave Propagation in Elastic Solids* (North-Holland, Amsterdam).

Anderson, D. M., Cermelli, P., Fried, E., Gurtin, M. E. and McFadden, G. B. (2007). General dynamical sharp-interface conditions for phase transformations in viscous heat-conducting fluids, *J. Fluid Mech.* **581**, pp. 323–370.

Atkinson, C. and Eshelby, J. D. (1968). The flow of energy into the tip of a moving crack, *Int. J. Fracture* **4**, pp. 3–8.

Bale, D. S., LeVeque, R. J., Mitran, S. and Rossmanith, J. A. (2003). A wave propagation method for conservation laws and balance laws with spatially varying flux functions. *SIAM J. Sci. Comp.* **24**, pp. 955–978.

Banks-Sills, S., Elasi, R. and Berlin, Y. (2002). Modeling of functionally graded materials in dynamic analyses, *Composites: Part B* **33**, pp. 7–15.

Barenblatt, G.I. (1996). *Scaling, Self-Similarity, and Intermediate Asymptotics* (Cambridge University Press).

Bedford, A. and Drumheller, D.S. (1994). *Introduction to Elastic Wave Propagation* (Wiley, New York).

Bekker, A., Jimenez-Victory, J. C., Popov, P. and Lagoudas, D. C. (2002). Impact induced propagation of phase transformation in a shape memory alloy rod, *Int. J. Plasticity* **18**, pp. 1447–1479.

Berezovski, A., Berezovski, M. and Engelbrecht, J. (2006). Numerical simulation of nonlinear elastic wave propagation in piecewise homogeneous media, *Mater. Sci. Engng.* **A 418**, pp. 364-369.

Berezovski, A., Engelbrecht, J. and Maugin, G. A. (2000). Thermoelastic wave propagation in inhomogeneous media, *Arch. Appl. Mech.* **70**, pp. 694–706.

Berezovski, A., Engelbrecht, J. and Maugin, G. A. (2003). Numerical simulation of two-dimensional wave propagation in functionally graded materials, *Eur. J. Mech. - A/Solids* **22**, pp. 257–265.

Berezovski, A. and Maugin, G. A. (2001). Simulation of thermoelastic wave propagation by means of a composite wave-propagation algorithm, *J. Comp. Physics*, **168**, pp. 249–264.

Berezovski, A. and Maugin, G. A. (2002). Thermoelastic wave and front propagation, *J. Thermal Stresses* **25**, pp. 719–743.

Berezovski, A. and Maugin, G. A. (2004). On the thermodynamic conditions at moving phase-transition fronts in thermoelastic solids, *J. Non-Equilib. Thermodyn.* **29**, pp. 37–51.

Berezovski, A. and Maugin, G. A. (2005). On the velocity of a moving phase boundary in solids, *Acta Mech.* **179**, pp. 187–196.

Berezovski, A. and Maugin, G. A. (2005). Stress-induced phase-transition front propagation in thermoelastic solids, *Eur. J. Mech. - A/Solids* **24**, pp. 1–21.

Bernardini, D. (2001). On the macroscopic free energy functions of shape memory alloys, *J. Mech. Phys. Solids* **49**, pp. 813–837.

Bernardini, D. and Pence, T. J. (2002). Shape memory materials: modeling, in M. Schwartz, (ed.), *Encyclopedia of Smart Materials* (Wiley, New York), pp. 964–980.

Billingham, J. and King, A. C. (2000). *Wave Motion* (Cambridge University Press).

Birman, V. (1997). Review of mechanics of shape memory alloy structures, *Appl. Mech. Rev.* **50**, pp. 629–645.

Boley, B. A. and Weiner, J. H. (1960). *Thermoelasticity* (Wiley, New York).

Bruck, H. A. (2000). A one-dimensional model for designing functionally graded materials to manage stress waves, *Int. J. Solids Struct.* **37**, pp. 6383–6395.

Callen, H. B. (1960). *Thermodynamics* (Wiley & Sons, New York).

Carlson, D. E. (1972). Linear thermoelasticity, in C.A. Truesdell (ed.), *Handbuch der Physik*, Vol. VI/2, (Springer, Berlin), pp. 297–345.

Casas-Vázquez, J. and Jou, D. (2003). Temperature in non-equilibrium states: a review of open problems and current proposals, *Rep. Prog. Phys.* **66**, pp. 1937–2023.

Cermelli, P. and Sellers, S. (2000). Multi-phase equilibrium of crystalline solids, *J. Mech. Phys. Solids* **48**, pp. 765–796.

Chakraborty, A. and Gopalakrishnan, S. (2003). Various numerical techniques for analysis of longitudinal wave propagation in inhomogeneous one-dimensional waveguides, *Acta Mech.* **162**, pp. 1–27.

Chakraborty, A. and Gopalakrishnan, S. (2004). Wave propagation in inhomogeneous layered media: solution of forward and inverse problems, *Acta Mech.* **169**, pp. 153–185.

Chen, X. and Chandra, N. (2004). The effect of heterogeneity on plane wave propagation through layered composites, *Comp. Sci. Technol.* **64**, pp. 1477–1493.

Chen, X., Chandra, N. and Rajendran, A. M. (2004). Analytical solution to the plate impact problem of layered heterogeneous material systems, *Int. J. Solids Struct.* **41**, pp. 4635–4659.

Chen, Y.-C. and Lagoudas, D. C. (2000). Impact induced phase transformation in shape memory alloys, *J. Mech. Phys. Solids* **48**, pp. 275–300.

Chiu, T.-C. and Erdogan, F. (1999). One-dimensional wave propagation in a functionally graded elastic medium, *J. Sound Vibr.* **222**, pp. 453–487.

Cho, J. R. and Ha, D. Y. (2001). Averaging and finite-element discretization approaches in the numerical analysis of functionally graded materials, *Mater. Sci. Engng. A* **302**, pp. 187–196.

Cho, J. R. and Oden, J. T. (2000). Functionally graded material: a parametric study on thermal-stress characteristics using the Crank-Nicolson-Galerkin scheme, *Comput. Methods Appl. Mech. Engng.* **188**, pp. 17–38.

Christian, J. W. (1965). *The Theory of Transformations in Metals and Alloys* (Pergamon, London).

Chrysochoos, A. Licht, C. and Peyroux, R. (2003). A one-dimensional thermomechanical modeling of phase change front propagation in a SMA monocrystal, *C. R. Mecanique* **331**, pp. 25–32.

Cotterell, B. (2002). The past, present, and future of fracture mechanics, *Engng. Fract. Mech.* **69**, pp. 533–553.

Cox, B. N., Gao, H., Gross, D. and Rittel, D. (2005). Modern topics and challenges in dynamic fracture, *J. Mech. Phys. Solids* **53**, pp. 565–596.

Dafermos, C. M. (2005). *Hyperbolic Conservation Laws in Continuum Physics*, 2nd Edition, (Springer, Berlin).

Dally, J. W. (1979). Dynamic photoelastic studies of fracture, *Exp. Mech.* **19**, pp. 349–361.

Dascalu, C. and Maugin, G. A. (1993). Material forces and energy-release rate in homogeneous elastic solids with defects, *C.R. Acad. Sci. Paris II* **317**, pp.

1135–1140.

Dascalu, C. and Maugin, G. A. (1995). The thermoelastic material-momentum equation, *J. Elasticity* **39**, pp. 201–212.

De Groot, S. R. and Mazur, P. (1961). *Non-Equilibrium Thermodynamics* (North-Holland, Amsterdam).

Duarte, C. A., Hamzeh, O. N., Liszka, T. J. and Tworzydlo, W. W. (2001). A generalized finite element method for the simulation of threedimensional dynamic crack propagation, *Comp. Meth. Appl. Mech. Engng.* **190**, pp. 2227–2262.

Emel'yanov, Y., Golyandin, S., Kobelev, N. P., Kustov, S., Nikanorov, S., Pugachev, G., Sapozhnikov, K., Sinani, A., Soifer, Ya. M., Van Humbeeck, J. and De Batist, R. (2000). Detection of shock-wave-induced internal stresses in Cu-Al-Ni shape memory alloy by means of acoustic technique, *Scripta mater.* **43**, pp. 1051–1057.

Engelbrecht, J., Berezovski, A., Pastrone, F. and Braun, M. (2005). Waves in microstructured materials and dispersion, *Phil. Mag.* **85**, pp. 4127–4141.

Epstein, M. and Maugin, G. A. (1995). Thermoelastic material forces: definition and geometric aspects, *C. R. Acad. Sci. Paris II* **320**, pp. 63–68.

Ericksen, J. L. (1977). Special topics in elastostatics, in: C.-S. Yih (ed.), *Advances in Applied Mechanics*, Vol. 17, (Academic Press, New York), pp. 189–244.

Ericksen, J. L. (1998). *Introduction to the Thermodynamics of solids* (Springer, New York).

Escobar, J. C. and Clifton, R. J. (1993). On pressure-shear plate impact for studying the kinetics of stress-induced phase-transformations, *Mat. Sci. & Engng.* **A170**, pp. 125–142.

Escobar, J. C. and Clifton, R. J. (1995). Pressure-shear impact-induced phase transformations in Cu-14.44Al-4.19Ni single crystals, in: *Active Materials and Smart Structures* (SPIE Proceedings), **2427**, pp. 186–197.

Evora, V. M .F., Jain, N, Shukla, A. (2005.) Stress intensity factor and crack velocity relationship for Polyester/TiO2 nanocomposites, *Exp Mech* **45**, pp. 153–159.

Fineberg, J. and Marder, M. (1999). Instability in dynamic fracture, *Phys. Reports* **313**, pp. 1-108.

Fischer, F. D., Berveiller, M., Tanaka, K. and Oberaigner, E. R. (1994). Continuum mechanical aspects of phase transformations in solids, *Arch. Appl. Mech.* **64**, pp. 54–85.

Fischer, F. D. and Simha, N. K. (2004). Influence of material flux on the jump relations at a singular interface in a multicomponent solid, *Acta Mech.* **171**, pp. 213–223.

Fogarthy, T. and LeVeque, R. J. (1999). High-resolution finite-volume methods for acoustics in periodic and random media, *J. Acoust. Soc. Am.* **106**, pp. 261–297.

Freund, L. B. (1972). Energy flux into the tip of an extending crack in an elastic solid, *J. Elasticity*, **2**, pp. 341–349.

Freund, L. B. (1990). *Dynamic Fracture Mechanics* (Cambridge University Press).

Gasik, M. M. (1998). Micromechanical modelling of functionally graded materials,

Comput. Mater. Sci. **13**, pp. 42–55.

Godlewski, E. and Raviart, P.-A. (1996). *Numerical Approximation of Hyperbolic Systems of Conservation Laws* (Springer, New York).

Goo, B. C. and Lexcellent, C. (1997). Micromechnics-based modeling of two-way memory effect of a single-crystalline shape-memory alloy, *Acta Mater.* **45**, pp. 727–737.

Grady, D. (1998). Scattering as a mechanism for structured shock waves in metals, *J. Mech. Phys. Solids* **46**, pp. 2017–2032.

Graff, K. F. (1975). *Wave Motion in Elastic Solids* (Oxford University Press).

Guinot, V. (2003). *Godunov-type Schemes: An Introduction for Engineers* (Elsevier, Amsterdam).

Gurtin, M. E. and Jabbour, M. E. (2002). Interface evolution in three dimensions with curvature-dependent energy and surface diffusion: interface-controlled evolution, phase transitions, epitaxial growth of elastic films, *Arch. Rat. Mech. Anal.* **163**, pp. 171–208.

Gurtin, M. E. and Voorhees, P. W. (1996). The thermodynamics of evolving interfaces far from equilibrium, *Acta Mater.* **44**, pp. 235–247.

Han, X., Liu, G. R. and Lam, K. Y. (2001). Transient waves in plates of functionally graded materials, *Int. J. Numer. Methods Engng.* **52**, pp. 851–865.

Hashin, Z. (1983). Analysis of composite materials - a survey, *J. Appl. Mech.* **50**, pp. 481–505.

Hauch, J. A. and Marder, M. P. (1998). Energy balance in dynamic fracture, investigated by a potential drop technique, *Int. J. Fracture* **90**, pp. 133–151.

Helm, D. and Haupt, P. (2003). Shape memory behaviour: modelling within continuum thermomechanics, *Int. J. Solids Struct.* **40**, pp. 827–849.

Hetnarski, R. B. (ed.), (1986). *Thermal Stresses*, Vol. I, (Elsevier, Amsterdam).

Hirai, T. (1996). Functionally graded materials, in: *Processing of Ceramics*, Vol. 17B, Part 2, (VCH Verlagsgesellschaft, Weinheim, Germany), pp. 292–341.

Hoffmann, K. H., Burzler, J. M. and Schubert, S. (1997). Endoreversible thermodynamics, *J. Non-Equil. Thermodyn.* **22**, pp. 311–355.

Iadicola, M. A. and Shaw, J. A. (2004). Rate and thermal sensitivities of unstable transformation behavior in a shape memory alloy, *Int. J. Plasticity* **20**, pp. 577–605.

Irschik, H. (2003). On the necessity of surface growth terms for the consistency of jump relations at a singular surface, *Acta Mech.* **162**, pp. 195–211.

Jou, D., Casas-Vazquez, J. and Lebon, G. (2001). *Extended Irreversible Thermodynamics*, 3rd edition, (Springer, Berlin).

Karihaloo, B. L. and Xiao, Q. Z. (2003). Modelling of stationary and growing cracks in FE framework without remeshing: a state-of-the-art review, *Comp. Struct.* **81**, pp. 119–129.

Keller, J. U. (1971). Ein Beitrag zur Thermodynamik fluider Systeme, *Physica* **53**, pp. 602–620.

Kestin, J. (1992). Local-equilibrium formalism applied to mechanics of solids, *Int. J. Solids Struct.* **29**, pp. 1827–1836.

Kestin, J. (1993). Internal variables in the local-equilibrium approximation, *J.*

Non-Equilib. Thermodyn. **18**, pp. 360–379.

Kienzler, R. and Herrmann, G. (2000). *Mechanics in Material Space* (Springer, Berlin).

Kim, S. J. and Abeyaratne, R. (1995). On the effect of the heat generated during a stress-induced thermoelastic phase transformation, *Continuum Mech. Thermodyn.* **7**, pp. 311–332.

Klein, P. A. and Gao, H. (1998). Crack nucleation and growth as strain localization in a virtual-bond continuum, *Engng. Fract. Mech.* **61**, pp. 21–48.

Kobayashi, A. S. and Mall, S. (1978). Dynamic fracture toughness of Homalite-100, *Exp. Mech.*,**18**, pp. 11–18.

Kostrov, B. V. and Nikitin, L. V. (1970). Some general problems of mechanics of brittle fracture, *Arch. Mech. Stosow.* **22**, pp. 749–775.

Kubair, D. V. and Geubelle, P. (2003). Comparative analysis of extrinsic and intrinsic cohesive models of dynamic fracture, *Int. J. Solids Struct.* **40**, pp. 3853–3868.

Lagoudas, D. C. , Ravi-Chandar, K., Sarh, K. and Popov P. (2003). Dynamic loading of polycrystalline shape memory alloy rods, *Mech. Mater.* **35**, pp. 689–716.

Landau, L. D. and Lifshitz, E. M. (1986). *Theory of Fields* (Pergamon, Oxford).

Langseth, J. O. and LeVeque, R. J. (2000). A wave propagation method for three-dimensional hyperbolic conservation laws, *J. Comp. Physics* **165**, pp. 126–166.

Lax, P. D. and Wendroff, B. (1964). Difference schemes for hyperbolic equations with high order of accuracy, *Comm. Pure Appl. Math.* **18**, pp. 381–398.

Lee, Y. and Prakash, V. (1999). Dynamic brittle fracture of high strength structural steels under conditions of plane strain, *Int. J. Solids Struct.* **36**, pp. 3293–3337.

LeFloch, P. G. (2002). *Hyperbolic Systems of Conservation Laws* (Birkhäuser, ETH Zürich).

Leggoe, J. W., Mammoli, A. A., Bush, M. B. and Hu, X. Z. (1998). Finite element modelling of deformation in particulate reinforced metal matrix composites with random local microstructure variation, *Acta mater.* **46**, pp. 6075–6088.

Leo, P. H., Shield, T. W. and Bruno, O. (1993). Transient heat transfer effects on the behavior of shape-memory wires, *Acta Metall. Mater.* **41**, pp. 2477–2485.

LeVeque, R. J. (1997). Wave propagation algorithms for multidimensional hyperbolic systems, *J. Comp. Phys.* **131**, pp. 327–353.

LeVeque, R. J. (1998). Balancing source terms and flux gradients in high-resolution Godunov methods: the quasi-steady wave-propagation algorithm, *J. Comp. Phys.* **148**, pp. 346–365.

LeVeque, R. J. (2002a). *Finite Volume Methods for Hyperbolic Problems* (Cambridge University Press).

LeVeque, R. J. (2002b). Finite volume methods for nonlinear elasticity in heterogeneous media, *Int. J. Numer. Methods Fluids* **40**, pp. 93–104.

LeVeque, R. J. and Yong, D. H. (2003). Solitary waves in layered nonlinear media, *SIAM J. Appl. Math.* **63**, pp. 1539–1560.

Li, Y. and Ramesh, K. T. (1998). Influence of particle volume fraction, shape, and aspect ratio on the behavior of particle-reinforced metal-matrix composites at high rates of strain, *Acta Mater.* **46**, pp. 5633–5646.

Li, Y., Ramesh, K. T. and Chin, E. S. C. (2001). Dynamic characterization of layered and graded structures under impulsive loading, *Int. J. Solids Struct.* **38**, pp. 6045–6061.

Liska, R. and Wendroff, B. (1998). Composite schemes for conservation laws, *SIAM J. Numer. Anal.* **35**, pp. 2250–2271.

Liu, G. R., Han, X. and Lam, K. Y. (1999). Stress waves in functionally gradient materials and its use for material characterization, *Composites, Part B* **30**, pp. 383–394.

Marder, M. and Gross, S. P. (1995). Origin of crack tip instabilities, *J. Mech. Phys. Solids* **43**, pp. 1–48.

Markworth, A. J., Ramesh, K. T. and Parks, W. P. (1995). Modelling studies applied to functionally graded materials, *J. Mater. Sci.* **30**, pp. 2183–2193.

Maugin, G. A. (1988). *Continuum Mechanics of Electromagnetic Solids* (North-Holland, Amsterdam).

Maugin, G. A. (1990). Internal variables and dissipative structures. *J. Non-Equilib. Thermodyn.* **15**, pp. 173–192.

Maugin, G. A. (1992). *Thermomechanics of Plasticity and Fracture* (Cambridge University Press).

Maugin, G. A. (1993). *Material Inhomogeneities in Elasticity* (Chapman and Hall, London).

Maugin, G. A. (1994). On the J-integral and energy-release rates in dynamical fracture, *Acta Mech.* **105**, pp. 33–47.

Maugin, G. A. (1995). Material forces: concepts and applications, *A.S.M.E., Appl. Mech. Rev.* **48**, pp. 213–245.

Maugin, G. A. (1996). On Ericksen-Noether identity and material balance laws in thermoelasticity and akin phenomena, in R.C. Batra and M.F. Beatty (eds.), *Contemporary Research in the Mechanics and Mathematics of Materials* (J.L. Ericksens 70th Anniversary Volume), (C.I.M.N.E., Barcelona), pp. 397–407.

Maugin, G. A. (1997). Thermomechanics of inhomogeneous-heterogeneous systems: application to the irreversible progress of two- and three-dimensional defects, *ARI* **50**, pp. 41–56.

Maugin, G. A. (1998). On shock waves and phase-transition fronts in continua. *ARI,* **50**, pp. 141–150.

Maugin, G. A. (1998-1999). Multiscale approach to a basic problem of materials mechanics (Propagation of phase-transition fronts), *Atti Accad. Pelor. Pericol. Fis. Mat. Nat.* **76-77**, pp. 169–190.

Maugin, G. A. (1999a). *Nonlinear Waves in Elastic Crystals* (Oxford University Press).

Maugin, G. A. (1999b). *The Thermomechanics of Nonlinear Irreversible Behavior. An Introduction* (World Scientific, Singapore).

Maugin, G. A. (2000). On the universality of the thermomechanics of forces driving singular sets, *Arch. Appl. Mech.,* **70**, pp. 31–45.

Maugin, G. A. (2003). Pseudo-plasticity and pseudo-inhomogeneity effects in materials mechanics, *J. Elasticity* **71**, pp. 81–103.

Maugin, G. A. (2006). On the thermomechanics of continuous media with diffusion and/or weak nonlocality, *Arch. Appl. Mech.* **75**, pp. 723–738.

Maugin, G. A. and Berezovski, A. (1999). Material formulation of finite strain thermoelasticity and applications, *J. Thermal Stresses* **22**, pp. 421–449.

Maugin, G. A. and Muschik, W. (1994). Thermodynamics with internal variables. Part 1. General concepts, *J. Non-Equilib. Thermodyn.* **19**, pp. 217–249.

Maugin, G. A. and Trimarco, C. (1995). The dynamics of configurational forces at phase-transition fronts, *Meccanica* **30**, pp. 605–619.

McKelvey, A. L. and Ritchie, R. O. (2000). On the temperature dependence of the superelastic strength and the prediction of the theoretical uniaxial transformation strain in Nitinol, *Phil. Mag. A* **80**, pp. 1759–1768.

Messner, C. and Werner, E. A. (2003). Temperature distribution due to localised martensitic transformation in SMA tensile test specimens, *Comput. Mater. Sci.* **26**, pp. 95–101.

Meurer, T., Qu, J. and Jacobs, L. J. (2002). Wave propagation in nonlinear and hysteretic mediaa numerical study, *Int. J. Solids Struct.* **39**, pp. 5585–5614.

Moran, B. and Shih, C. F. (1987a). A general treatment of crack tip contour integrals, *Int. J. Fracture* **35**, pp. 295–310.

Moran, B. and Shih, C. F. (1987b). Crack tip and associated domain integrals from momentum and energy balance, *Engng. Fract. Mech.* **27**, pp. 615–642.

Muschik, W. (1977). Empirical foundation and axiomatic treatment of non-equilibrium temperature, *Arch. Rat. Mech. Anal.* **66**, pp. 379–401.

Muschik, W. (1979). Fundamentals of dissipation inequalities, *J. Non-Equilib. Thermodyn.* **4**, pp. 277–294.

Muschik, W. (1980). Entropies of heat conducting discrete or multi-temperature systems with use of non-equilibrium temperatures, *Int. J. Engng. Sci.*, **18**, pp. 1399–1410.

Muschik, W. (1984). Fundamental remarks on evaluating dissipation inequalities, in J. Casas-Vazquez, D. Jou, G. Lebon, (eds), *Recent Developments in Nonequilibrium Thermodynamics* (Springer, Berlin), pp. 388–397.

Muschik, W. (1990). *Aspects of Non-Equilibrium Thermodynamics* (World Scientific, Singapore).

Muschik, W. (1993). Fundamentals of non-equilibrium thermodynamics, in W. Muschik, (ed.), *Non-Equilibrium Thermodynamics with Application to Solids* (Springer, Wien), pp. 1–63.

Muschik, W. and Berezovski, A. (2004). Thermodynamic interaction between two discrete systems in non-equilibrium, *J. Non-Equilib. Thermodyn.* **29**, pp. 237–255.

Muschik, W. and Berezovski, A. (2007). Non-equilibrium contact quantities and compound deficiency at interfaces between discrete systems, *Proc. Estonian Acad. Sci. Phys. Math.* **56**, pp. 133-145.

Muschik, W. and Brunk, G. (1977). A concept of non-equilibrium temperature, *Int. J. Engng. Sci.* **15**, pp. 377–389.

Muschik, W. and Gümbel, S. (1999). Does Clausius' inequality exists for open systems? *J. Non-Equilib. Thermodyn.* **24**, pp. 97–106.

Muschik, W., Papenfuss, C. and Ehrentraut, H. (1997). *Concepts of Continuum Thermodynamics* (Technische Universität Berlin, Kielce University of Technology, Kielce, Poland).

Muschik, W., Papenfuss, C. and Ehrentraut, H. (2001). A sketch of continuum thermodynamics, *J. Non-Newtonian Fluid Mech.* **96**, pp. 255–290.

Nakamura, T., Shih, C. F. and Freund, L. B. (1985). Computational methods based on an energy integral in dynamic fracture, *Int. J. Fracture* **27**, pp. 229–243.

Needleman, A. (1997). Numerical modeling of crack growth under dynamic loading conditions, *Comp. Mech.* **19**, pp. 463–469.

Ngan, S.-C., Truskinovsky, L. (2002). Thermo-elastic aspects of dynamic nucleation, *J. Mech. Phys. Solids* **50**, pp. 1193–1229.

Nishioka, T. (1997). Computational dynamic fracture mechanics, *Int. J. Fracture* **86**, pp. 127–159.

Nowacki, W. (1986). *Thermoelasticity*, 2nd Edition, (Pergamon Press, Oxford and P.W.N., Warsaw).

O'Reilly, O. M. and Varadi, P. C. (1999). A treatment of shocks in one-dimensional thermomechanical media, *Continuum Mech. Thermodyn.* **11**, pp. 339–352.

Orgeas, L. and Favier, D. (1998). Stress-induced martensitic transformation of a NiTi alloy in anisothermal shear, tension and compression, *Acta Mater.* **46**, pp. 5579–5591.

Purohit, P. K. and Bhattacharya, K. (2003). Dynamics of strings made of phase-transforming materials, *J. Mech. Phys. Solids* **51**, pp. 393–424.

Qidwai, M. A. and Lagoudas, D. C. (2000). On thermomechanics and transformation surfaces of polycrystalline NiTi shape memory alloy material, *Int. J. Plasticity* **16**, pp. 1309–1343.

Ravi-Chandar, K. (1998). Dynamic fracture of nominally brittle materials, *Int. J. Fracture* **90**, pp. 83–102.

Ravi-Chandar, K. (2004). *Dynamic Fracture* (Elsevier, Amsterdam).

Ravi-Chandar, K. and Knauss, W. G. (1984a). An experimental investigation into dynamic fracture: I. crack initiation and arrest, *Int. J. Fracture* **25**, pp. 247–262.

Ravi-Chandar, K. and Knauss, W. G. (1984b). An experimental investigation into dynamic fracture: II. microstructural aspects, *Int. J. Fracture* **26**, pp. 65–80.

Ravi-Chandar, K. and Knauss, W. G. (1984c). An experimental investigation into dynamic fracture: III. On steady state crack propagation and crack branching, *Int. J. Fracture* **26**, pp. 141–154.

Ravi-Chandar, K. and Knauss, W. G. (1984d). An experimental investigation into dynamic fracture: IV. On the interaction of stress waves with propagating cracks, *Int. J. Fracture* **26**, pp. 189–200.

Ravi-Chandar, K. and Knauss, W. G. (1985). Some basic problems in stress wave dominate fracture, *Int. J. Fracture* **27**, pp. 127–144.

Ravichandran, G. and Clifton, R. J. (1989). Dynamic fracture under plane wave loading, *Int. J. Fracture* **40**, pp. 157–201.

Réthoré, J., Gravouil, A. and Combescure, A. (2004). A stable numerical scheme for the finite element simulation of dynamic crack propagation with remeshing, *Comp. Meth. Appl. Mech. Engng.* **193**, pp. 4493–4510.

Rice, J. R. (1968). A path independent integral and the approximate analysis of strain concentration by notches and cracks, *ASME J. Appl. Mech.* **35**, pp. 379–386.

Rokhlin, S. I. and Wang, L. (2002). Ultrasonic waves in layered anisotropic media: characterization of multidirectional composites, *Int. J. Solids Struct.* **39**, pp. 5529–5545.

Santosa, F. and Symes, W. W. (1991). A dispersive effective medium for wave propagation in periodic composites, *SIAM J. Appl. Math.* **51**, pp. 984–1005.

Schottky, W. (1929). *Thermodynamik* (Springer, Berlin).

Sharon, E. and Fineberg, J. (1999). Confirming the continuum theory of dynamic brittle fracture for fast cracks. *Nature* **397**, pp. 333–335.

Shaw, J. A. (2000). Simulations of localized thermo-mechanical behavior in a NiTi shape memory alloy, *Int. J. Plasticity* **16**, pp. 541–562.

Shaw, J. A. (2002). A thermomechanical model for a 1-D shape memory alloy wire with propagating instabilities, *Int. J. Solids Struct.* **39**, pp. 1275–1305.

Shaw, J. A. and Kyriakides, S. (1995). Thermomechanical aspects of NiTi, *J. Mech. Phys. Solids* **43**, pp. 1243–1281.

Shaw, J. A. and Kyriakides, S. (1997). On the nucleation and propagation of phase transformation fronts in a NiTi alloy, *Acta Mater.* **45**, pp. 683–700.

Shih, C.F., Moran, B. and Nakamura, T. (1986). Energy release rate along a three-dimensional crack front in a thermally stressed body, *Int. J. Fracture* **30**, pp. 79–102.

Stoilov, V. and Bhattacharyya, A. (2002). A theoretical framework of one-dimensional sharp phase fronts in shape memory alloys, *Acta Mater.* **50**, pp. 4939–4952.

Suhubi, E. S. (1975). Thermoelastic solids, in A.C. Eringen (ed.), *Continuum Physics*, Vol. 2, (Academic Press, New York), pp. 174–265.

Sun, Q. P., Li, Z. Q. and Tse, K. K. (2000). On superelastic deformation of NiTi shape memory alloy micro-tubes and wiresband nucleation and propagation, in U. Gabbert, H.S. Tzou, (eds.), *Proceedings of IUTAM Symposium on Smart Structures and Structronic Systems* (Kluwer, Dordrecht), pp. 113–120.

Suresh, S. and Mortensen, A. (1998). *Fundamentals of Functionally Graded Materials* (The Institute of Materials, IOM Communications Ltd., London).

Toro, E. F. (1997). *Riemann Solvers and Numerical Methods for Fluid Dynamics* (Springer, Berlin).

Toro, E. F. (ed.) (2001). *Godunov Methods: Theory and Applications* (Kluwer, New York).

Truesdell, C. and Bharatha, S. (1977). *The Concepts and Logic of Classical Thermodynamics as a Theory of Heat Engines* (Springer, New York).

Truskinovsky, L. (1987). Dynamics of nonequilibrium phase boundaries in a heat conducting nonlinear elastic medium, *J. Appl. Math. Mech.* (PMM) **51**, pp. 777–784.

Vitiello, A., Giorleo, G. and Morace, R. E. (2005). Analysis of thermomechanical behaviour of Nitinol wires with high strain rates, *Smart Mater. Struct.* **14**, pp. 215–221.

Wang, L. and Rokhlin, S. I. (2004). Recursive geometric integrators for wave propagation in a functionally graded multilayered elastic medium, *J. Mech. Phys. Solids* **52**, pp. 2473–2506.

Wilmanski, K. (1988). *Thermomechanics of Continua* (Springer, Berlin).

Xu, X. P. and Needleman, A. (1994). Numerical simulations of fast crack growth in brittle solids, *J. Mech. Phys. Solids* **42**, pp. 1397–1437.

Zhong, X., Hou, T. Y. and LeFloch, P. G. (1996). Computational methods for propagating phase boundaries, *J. Comp. Physics* **124**, pp. 192–216.

Zhuang, S., Ravichandran, G. and Grady, D. (2003). An experimental investigation of shock wave propagation in periodically layered composites, *J. Mech. Phys. Solids* **51**, pp. 245–265.

Index

adiabatic case, 28, 29, 41, 97, 109, 112, 115, 117, 118, 120

balance of pseudomomentum, 15, 17, 21, 26, 47, 57
Boussinesq equation, 63
brittle fracture, 18, 22, 30

canonical equations, 18
Clausius-Duhem inequality, 13, 57
composite system, 34, 190, 197, 199, 201
compound deficiency, 197, 200, 205, 207
conservation laws, 1, 6, 9, 23, 49, 71, 85
constitutive relations, 1, 14, 40, 110, 189
contact quantities, 38, 39, 53, 190, 192, 193, 195, 200
crack front, 175, 176, 178, 179
crack tip, 3, 19, 22, 171, 174, 176

defining inequalities, 191
direct-motion gradient, 10, 16
discontinuity surface, 1, 12, 23, 108, 175
discrete systems, 6, 38, 39, 189, 190, 206
dispersion, 61, 68
dissipation rate, 21
driving force, 2, 5–7, 19, 23, 24, 26, 28–30, 41, 43, 57, 86, 94, 96–100, 103, 113, 115–119, 122, 172, 177, 183

effective media theory, 61
elastic wave propagation, 65, 140, 141, 143, 147
endoreversible system, 203, 204, 207
energy release rate, 3, 4, 19, 21, 171, 177
entropy production, 3, 4, 31, 41, 44, 48, 57, 117, 120, 122, 126, 177, 199, 203
equilibrium conditions, 34, 35
Ericksen's identity, 17
Eshelby stress tensor, 17, 22, 24, 26, 27, 47
excess quantities, 6, 39–41, 44, 53, 54, 56, 59, 90–92, 99, 108, 134, 145, 169, 197, 205, 207

field singularity, 2, 19, 30
finite volume algorithm, 7, 59, 89, 108, 129, 145, 157
fluctuation splitting, 130
free energy, 14–16, 25, 26, 29, 40, 45, 46, 53, 86, 89, 104, 110, 122, 175
front tracking, 100
functionally graded materials, 80–82, 147, 149, 152, 155, 157

Gibbs equation, 202

Godunov numerical scheme, 6, 51,
133, 135, 139

heat propagation, 13, 46
homothermal, 23, 27, 109, 113, 118
Hooke's law, 49, 60
hot heat source, 22

impact loading, 68, 70, 87, 103, 109,
147, 173
impedance mismatch, 70, 79
inhomogeneity force, 17, 47
initiation criterion, 93, 94, 99, 112,
125
integral balance laws, 11–13, 145
inverse-motion gradient, 10

J-integral, 4, 22, 177, 179
jump relations, 6, 9, 12, 13, 23, 28,
29, 31, 41, 44, 47, 54, 56, 86, 89, 90,
93, 97, 108, 116, 125, 161, 175, 176

kinematic compatibility, 60, 90, 92,
111
kinetic relation, 5, 6, 43, 44, 86, 88,
99, 117, 125, 180

latent heat, 110, 113, 114, 117, 120,
122, 126
Lax-Wendroff numerical scheme, 135,
139
Legendre transformation, 18, 27
linear elasticity, 60, 129, 171, 172, 176
linear waves, 59, 65
local balance laws, 12, 45
local equilibrium, 6, 34, 39, 53, 86,
168, 176, 178, 183, 203
local phase equilibrium, 31, 34, 36, 38

material force, 3, 15, 17, 21, 23, 48,
175
material inhomogeneities, 2, 9, 14, 23
Maxwell-Hadamard lemma, 25, 43
moving discontinuities, 1, 31, 41, 43,
44, 47, 85

moving phase boundary, 88, 89, 91,
108, 115, 120, 122, 126
multilayered FGM model, 149–151,
153, 155, 157

non-equilibrium, 34, 38, 39, 57, 97,
99, 190, 195, 202
nonlinear stress-strain relation, 71
nonlinear waves, 59, 62, 68
numerical fluxes, 6, 44, 51, 53, 54, 57,
100, 128, 129

periodic media, 60, 62
phase boundary, 5, 41, 86, 92–94, 96,
98, 99, 101, 102, 106, 112, 115, 116,
122, 125, 161, 162, 164
phase-transition front, 1, 4, 7, 15, 23,
59, 85, 99, 108, 116, 120, 125, 164,
169
Piola-Kirchhoff formulation, 9
Piola-Kirchhoff stress tensor, 11, 26,
27, 43
pseudoelasticity, 86, 102
pseudomomentum, 5, 15, 17, 18, 23,
24, 29, 47, 57

quasi-inhomogeneities, 2, 14, 44

randomly embedded particles, 150,
151, 154, 157
Rankine-Hugoniot condition, 13, 51
Rayleigh wave velocity, 171, 178
replacement quantities, 195
Riemann invariants, 56
Riemann problem, 51, 52, 55, 127,
134
rule of mixtures, 150

Schottky system, 190, 197
shape-memory alloy, 85, 87, 102, 103,
109, 122, 164
small-strain approximation, 45, 47
soliton-like pulses, 63
straight brittle crack, 3, 20, 171
stress-induced phase-transition front,
85, 109